MEM09002

2020

Interpret technical drawing

First Published November 2020

Conditions of Use:

Unit Resource Manual

Manufacturing Skills Australia Courses

This Student's Manual has been developed by BlackLine Design for use in the Manufacturing Skills Australia Courses.

Additional resource units can viewed and be ordered at www.acru.com.au

Feedback:

Your feedback is essential for improving the quality of these manuals.

This unit has not been technically edited. Please advise BlackLine Design of any changes, additions, deletions, or anything else you believe would improve the quality of this Student Workbook.   Don't assume that someone else will do it.   Your comments can be made by photocopying the relevant pages and including your comments or suggestions.

Forward your comments to:

>BlackLine Design
>
>blakline@bigpond.net.au
>
>Sydney, NSW   2000

Corporate Licenses

State and National TAFE Colleges and Institutes, and Registered Training Organisations are eligible to purchase corporate licenses.

All licenses are perpetual and allow the licensee to upload the material onto a delivery system (Moodle etc), print the resource in book form and sell or distribute the material to enrolled students within their organisation. The license allows the holder to re-badge the material but must retain acknowledgment to BlackLine Design as the original developer and owner.

Aims of the Competency Unit:

This unit of competency defines the skills and knowledge required to interpret technical drawings.

Technical drawings may utilise perspective, exploded views or hidden view techniques and may include symbol glossaries. Drawings are provided to AS 1100 Technical drawing or AS 1102 Graphical symbols and their equivalents from the full range of engineering disciplines.

Where any technical drawing, sketch, chart, diagram is only used as a technique for communication, then this unit does not apply: unit MEM12023 Perform engineering measurements or unit MEM16006 Organise and communicate information should be selected as appropriate.

Unit Hours:
36 Hours

Prerequisites:
MEM12023   Perform engineering measurements
MEM12024   Perform computations
MEM13015   Work safely and effectively in manufacturing and engineering
MEM16006   Organise and communicate information

Elements and Performance Criteria

1.	Determine job requirements	1.1	Follow standard operating procedures (SOPs)
		1.2	Comply with work health and safety (WHS) requirements at all times
		1.3	Identify job requirements from specifications, job sheets or associated work instructions
2.	Interpret technical drawing	2.1	Check drawing and version and validate against job requirements
		2.2	Recognise components and assemblies or objects
		2.3	Identify dimensions, instructions and material requirements
		2.4	Recognise symbols used in the drawing
		2.5	Compile list of required materials

Range of Conditions

Drawing interpretation includes recognising the following:

- relationship between the views contained in the drawing
- objects
- units of measurement
- dimensions of the key features
- symbols

Lesson Program:

Unit hour unit and is divided into the following program.

Topic	Skill Practice Exercise
Topic 1 – Engineering Drawings:	MEM09002-SP-0101
Topic 2 – Line Styles:	MEM09002-SP-0201 & MEM09002-SP-0202
Topic 3 – Reading Drawings:	MEM09002-SP-0301 & MEM09002-SP-0302
Topic 4 – Orthogonal Projection:	MEM09002-SP-0401 & MEM09002-SP-0402
Topic 5 – Units of Measurement:	MEM09002-SP-0501
Topic 6 – Assembly Drawings:	MEM09002-SP-0601 & MEM09002-SP-0602
Topic 7 – Abbreviations, Symbols & Notes:	MEM09002-SP-0701 & MEM09002-SP-0702
Practice Competency Test	MEM –PT-01

Contents:

Topic 1 – Engineering Drawings:

Required Skills:
- Name AS 1100 as the drawing standards setting out the basic principles of technical drawing practice.
- Identify the various types of drawings produced by drawing offices.

Required Knowledge:
- AS 1100 Drawing Standards.
- Development procedures for drawings.

Introduction:
The engineering drawing is one of the most important communication tools that a company can possess. Drawings are not only technical information, but also legal documents. Their creation and maintenance are expensive and time consuming. For these reasons, the effort made in fully understanding them cannot be taken for granted.

A drawing is one method of presenting technical communication. Technical communication is an advanced form of communication whereby people of the same trade (profession) can convey messages to one another more accurately and precisely. To achieve this, a technical language (and jargon), which is well standardized, is needed (e.g. botanical names in Horticulture and Latin for medical terminology, etc.).

Drawings have been used since the beginning of history for planning and producing art objects, architectural designs, and engineering works. Since the Industrial Revolution a system for creating architectural and engineering drawings has evolved. While the pens, pencils, tools, and papers for creating drawings have changed, the basic forms for presenting information have stayed the same. People producing technical drawings need to be familiar with the standard ways of presenting design information.

The ability to read and understand information contained on drawings is essential to perform most engineering-related jobs. Engineering drawings are the industry's means of communicating detailed and accurate information on how to manufacture/fabricate, assemble, troubleshoot, repair, and operate a piece of equipment or a system. To understand how to "read" a drawing it is necessary to be familiar with the standard conventions, rules, and basic symbols used on the various types of drawings. Before learning how to read the actual "drawing," an understanding of the information contained in the various non-drawing areas of a print is also necessary.

Draftspersons will inevitably be required to communicate with different people for different reasons as represented in Figure 1.1. In some situations, communications will be sufficiently taken care of by use of plain text. However, in other situations, text alone may not suffice and a more specialized form of communication by a technical engineering drawing may prove irreplaceably useful.

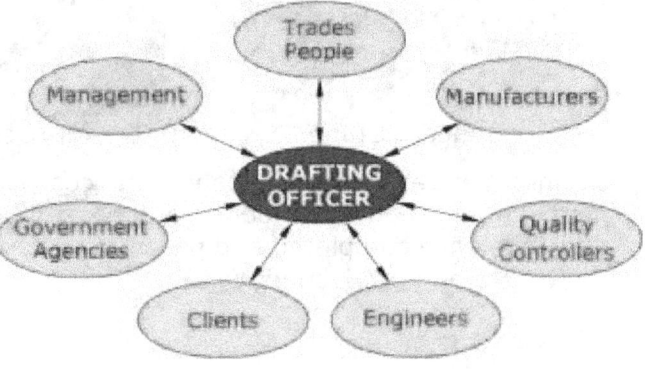

Figure 1. 1

Standards:

"Standardization is the process of formulating and applying rules for an orderly approach to a specific activity for the benefit and with the cooperation of all concerned, and in particular for the promotion of optimum overall economy taking due account of functional conditions and safety requirements." (International Organization for Standardization).

ISO or International Standards ensure that products and services are safe, reliable and of good quality. For business, they are strategic tools that reduce costs by minimizing waste and errors and increasing productivity. A standard is a document that provides requirements, specifications, guidelines or characteristics that can be used consistently to ensure that the materials, products, processes and services are fit for their purpose.

In Australia, AS1100 Australian Drawing Standards sets out the basic principles of technical drawing practice and covers:

The use of abbreviations.

- Materials, sizes, and layout of drawing sheets.
- The types and minimum thicknesses of lines to be used.
- The requirements for distinct uniform letters, numerals and symbols.
- Recommended scales and their application.
- Methods of projection and of indicating the various views of an object.
- Methods of sectioning.
- Recommendations for dimensioning including size and geometrical tolerancing.
- Conventions used for the representation of components and repetitive features of components.

Some the major industry disciplines include mechanical, automotive, architectural, civil and aeronautical.

Development of the Drawing:

Interpreting information from drawings is an important skill. Engineers and architects must be able to look at a set of plans and mentally picture the shapes of objects. Skilled workers must have the same abilities. Reading a drawing involves a highly developed ability to look at lines on the page and convert the shapes from several pictures to form a three-dimensional mental image. A product basically passes through three main stages; The Concept, then the Drawing Production, and finally manufacture; other stages such as estimating, costing and testing are involved but are supplementary to the main design and production.

Concept Stage

Design/Detail Stage

Manufacture Stage

During the Design/Detail stage, the components and assemblies are constantly being modified and redrawn or edited and passed between designer, detailer and engineer. The drawings once completed and passed, are forwarded to the workshops for manufacture by the trades and production workers.

Types of Drawings:

Drawing is one of the basic forms of visual communication and is used to record objects and actions of everyday life in an easily recognizable manner. There are two major types of drawings: artistic drawings and technical drawings.

Artistic drawings are a form of freehand representation that makes use of pictures to provide a general impression of the object being drawn. There are no hard rules or standards in the preparation of artistic drawings.

Artistic drawings are simply drawn by artists, based more or less on one's talent and skills. Although these drawings are often very attractive, they find very limited use in engineering disciplines.

Technical Drawings are detailed drawings drawn accurately and precisely; they are views of objects that have been prepared with the aid of computer programs or technical drawing/drafting instruments in order to record and transmit technical information. The drawings provide an exact and complete description of things that are to be built or manufactured.

- Technical drawings do not portray the objects the way they directly appear to the eye.
- They make use of many specialized symbols and conventions in order to transmit technical information clearly and exactly.
- To understand and correctly interpret technical drawings, one needs to acquaint oneself with the fundamentals of technical drawing; hence the purpose of this unit of competency.

The presentation of engineering or technical drawings is accomplished through several varying types of drawings including Freehand Sketches, Detail Drawings, Assembly Drawings, Pictorial, Schematic Diagrams and Circuit Diagrams.

Freehand Sketch:

Sketching is the creation of graphic images that are graphical representations or models of objects drawn in proportion but to no particular scale. Freehand sketching is manual sketching with the minimum of tools such as paper and pencil. Technical sketching is the art of creating a technical drawing using freehand without special instruments. Technical sketching requires correct shape or form and more so correct size indication. Generally, drawing tools refer to the materials used as aids when creating drawings and they vary from simple to complex instruments and equipment. However, modern drawing needs have changed dramatically due to the availability of computers. Traditional design and drafting has largely given way to computer design drafting, but design sketches will always be needed.

Sketches are helpful in capturing design ideas and trying out different solutions in a fast and inexpensive way; sketches are also useful for recording details of a job "on-site" which will be drawn correctly at a later date in the Drawing Office. Technical sketching is used as aid in conceptualization, spatial visualization and translating imagination into visual models. It could also be used as a means to amplify, clarify and record verbal explanations. Freehand sketching is an economic and effective means of formulating alternate solutions to a given problem so that a choice can be made on the best solution. Preliminary design studies are usually done with freehand sketches because accurate and detailed drawing of design options is expensive and time wasting at the initial stages of a project.

Artistic ability is an asset, but anyone can learn to sketch by following basic techniques. Draftspersons and Engineers frequently use special sketching grids which help keep lines straight and in proportion.

Words and notes on sketches must be readable and placed using uppercase characters to assure clarity. Cursive or script writing is never used as it is often unreadable after sketches and memos are duplicated, emailed or faxed to another location. Vertical capital block form letters are preferred.

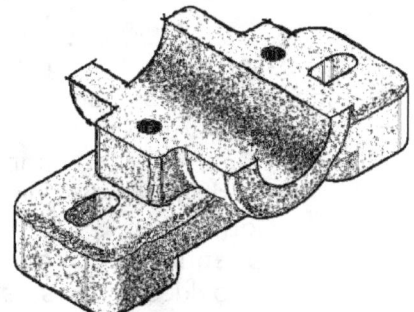

Figure 1. 2

Figure 1. 2 shows a freehand sketch of a Plumber Block Base; the sketch would normally include dimensions and notations but not the shading.

Detail Drawing:

A detail drawing is a print that shows a single component or part. It includes a complete and exact description of the part's shape and dimensions, and how it is made. A complete detail drawing will show in a direct and simple manner the shape, exact size, type of material, finish for each part, tolerance, necessary shop operations, number of parts required, and special notes for the manufacture or treatment after manufacture. A detail drawing is not the same as a detail view. A detail view shows part of a drawing in the same plane and in the same arrangement, but in greater detail to a larger scale than in the principal view.

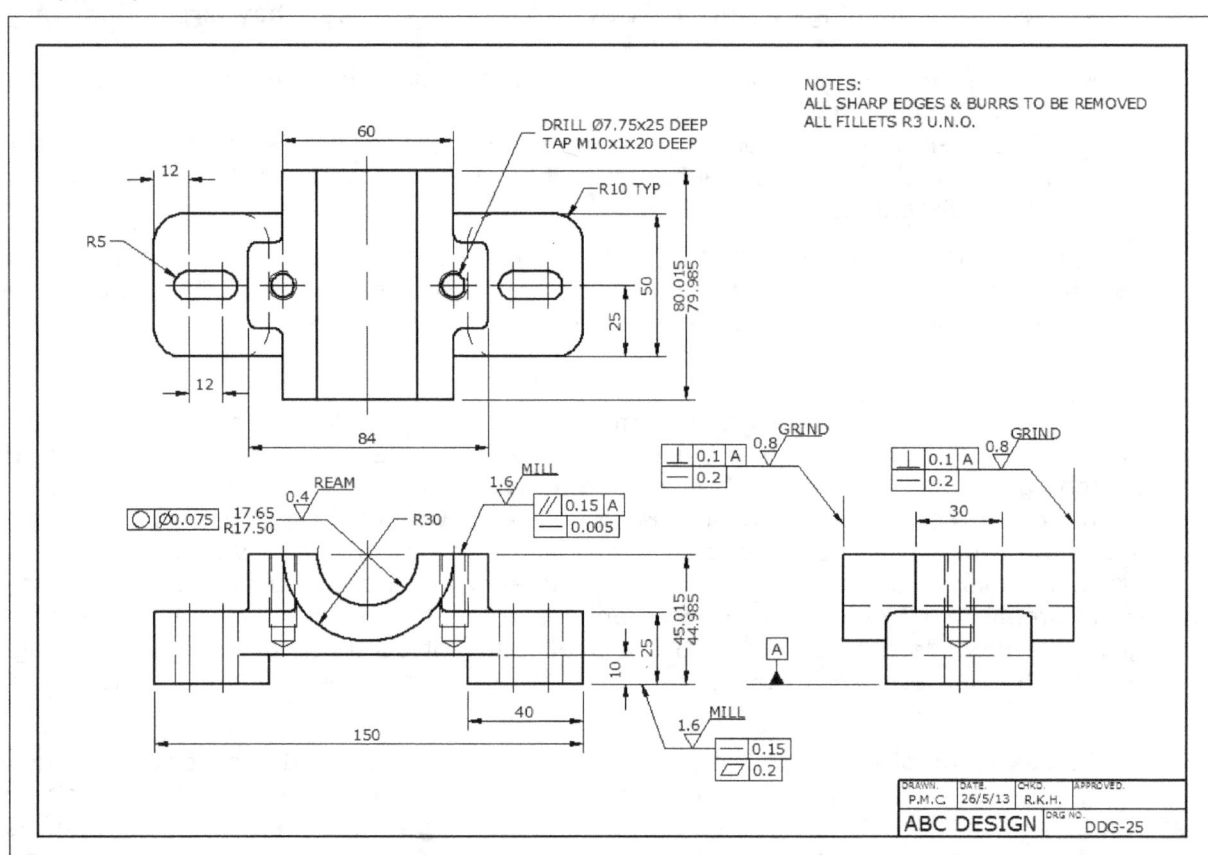

Figure 1.3

Figure 1.3 shows a detail drawing of the Plumber Block Base as sketched in Figure 1. 2. Three views have been provided to describe the shape while all dimensions, surface finish, general and geometric tolerances, and notations have been included on a completed drawing sheet.

Assembly Drawing:

An assembly working-drawing indicates how the individual parts of a machine or mechanism are assembled to make a complete unit. An assembly drawing serves the following purposes:

- Describes the shape of the assembled unit.
- Indicates how the parts of the assembled unit are positioned in relation to each other.
- Identifies each component that forms part of the assembled unit.
- Provides a parts list that describes and lists essential data concerning each part of the assembled unit.
- Provides, when necessary, reference information concerning the physical or functional characteristics of the assembled unit.

Assembly drawings may show one, two or three views to describe the assembled components; they must contain a Parts List (may also be called Material or Cutting List depending on the engineering discipline), cross-referencing (in balloons or circles), and General Notes pertaining to the assembly. The drawings normally show the over dimensions and centre-to-centre distances for specific assemblies.

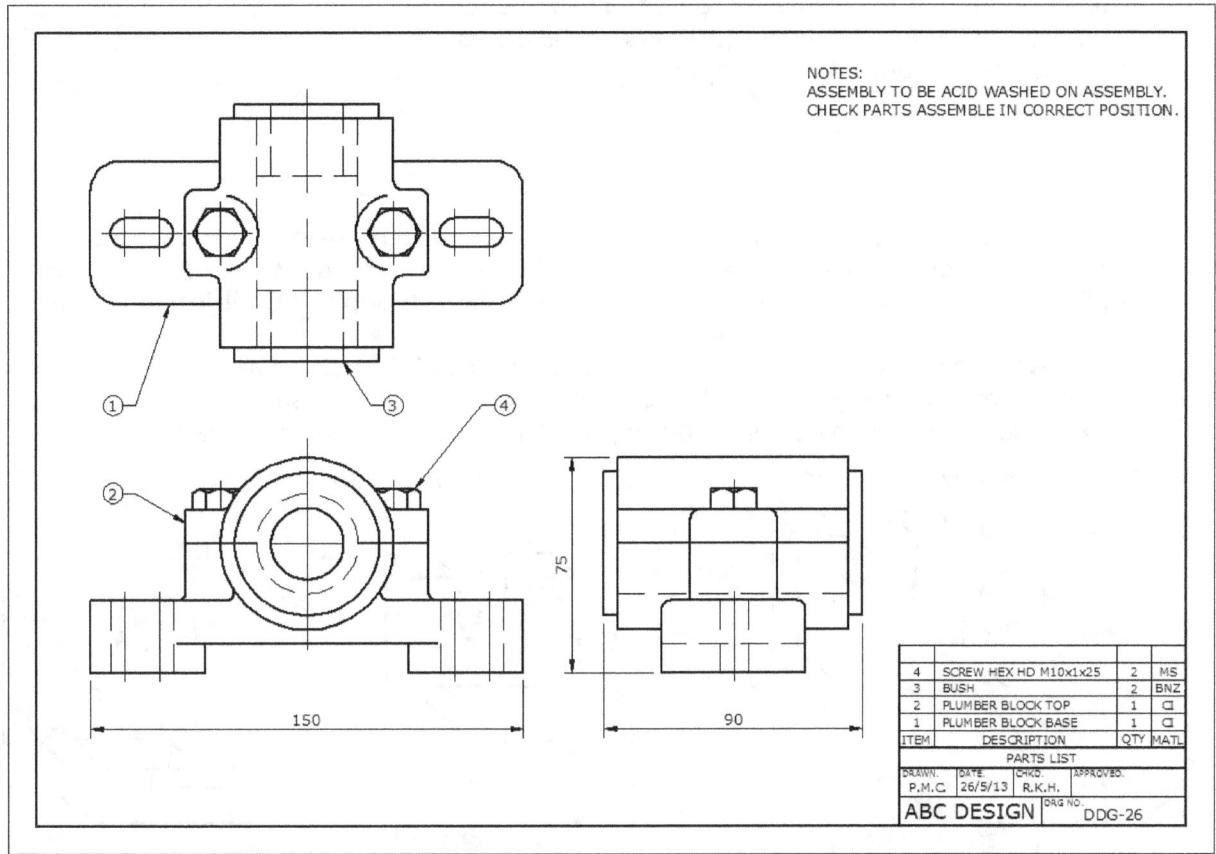

Figure 1.4

Figure 1.4 shows a completed assembly drawing with the Plumber Block Base, Plumber Block Top and Bushes drawn in place and secured with the Hexagonal Head Screws.

Most designs are commenced with an assembly drawing and when the concept of the design is finalised, the separate components can be broken out and detailed accordingly.

Pictorial:

Pictorial drawings are wrongly referred to as 3-D drawings. Pictorial drawings represent the shape of an object to show the three principal dimensions (length, width and height); it depicts the way people are used to viewing the object in everyday life but is drawn in 2-D. Characteristics of pictorial drawings are:

- The shapes are easier to visualise, and intersections of surfaces can be seen.
- Used for advertising, technical and repair manuals, and for general information.
- Pictorials can distort the lengths of lines and angles at corners; due the distortion factor, pictorial drawings are rarely used for production drawings.
- Pictorial drawings are 2-D drawings where the length along the Z-axis is 0 (zero).

The majority of pictorial drawings are produced as Isometric, Oblique, Axonometric or Perspective drawings.

Isometric:

Isometric drawings show three sides in dimensional proportion, but none are shown as a true shape with 90° corners. All the vertical lines are drawn vertically but all horizontal lines are drawn at 30° to the base line. All entities are drawn to scale. Circles and arcs are drawn as ellipses. Isometric is an easy method for presenting 3-D shapes.

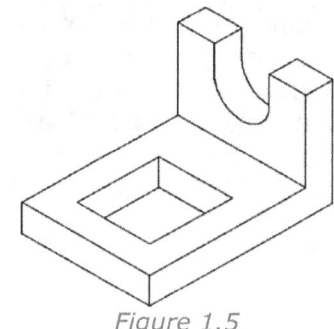

Figure 1.5

Oblique:

Oblique drawings are also designed to show a three-dimensional view of an object. The widths of the object are drawn as horizontal lines, but the depth is drawn back at a 45° angle. Three types of oblique drawings can be used to depict the object, normal, cavalier and cabinet obliques.

- Cavalier drawings display the depth using the full measurement.
- Normal drawings display the depth using ¾ of the measurement.
- Cabinet drawings display the depth using ½ of the measurement.

Circles are easier to draw in oblique as the circles can be drawn using a compass.

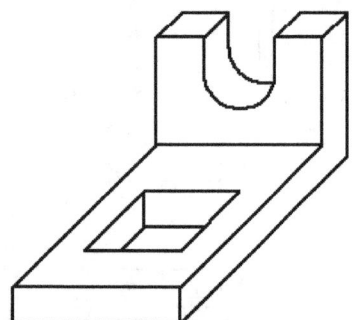
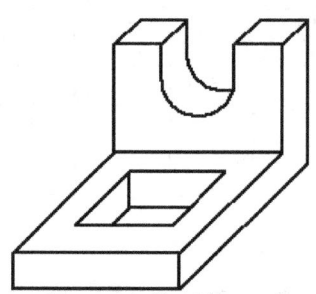
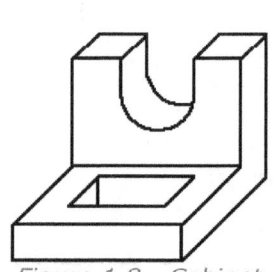

Figure 1.6 - Cavalier   *Figure 1.7 - Normal*   *Figure 1.8 - Cabinet*

Of the three images above only Figure 1.7 - Normal appears similar to the real-life object when viewed in oblique, Figure 1.6 - Cavalier appears too elongated while Figure 1.8 - Cabinet is too short or stubby.

Axonometric:

In Axonometric drawings, the object's vertical lines are drawn vertically, while the horizontal lines in the width and depth planes are shown at 30°-60° to the horizontal; in other words, the Plan or Top View is rotated through 30° or 60°.

Many kitchen manufacturers utilise axonometric in conveying the proposed arrangement of a new kitchen to a client.

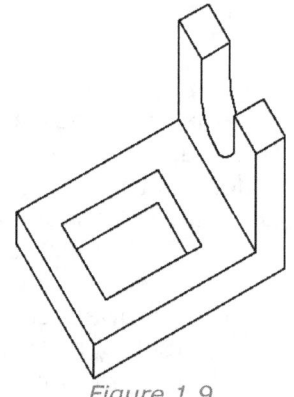

Figure 1.9

Perspective:

Perspective excels over all other types of projection in the pictorial representation of objects because it more closely approximates the view obtained by the human eye. Geometrically, a photograph is in perspective because the camera captures the same data an eye sees. Perspective is an important tool to designers, architects and engineers however is seldom used apart from architectural applications.

The elements required to produce perspective drawings can become quite daunting; these elements include Picture Planes, Station Points, Left & Right Vanishing Points and Horizon Plane.

Three types of perspective drawings are available, One-point, Two-point and Three Point Perspective.

One-Point Perspective:

In one-point perspective, the object is placed so that two sets of its principal edges are parallel to the Picture Plane, and the third set is perpendicular to the Picture Plane. The third set of parallel lines will converge toward a single vanishing point in perspective.

Two-Point Perspective:

In two-point perspective, the object is placed so that one set of parallel edges is vertical and has no vanishing point, while the other two sets each have vanishing points. Two-point perspective is the most common type used and is especially suitable for displaying houses and large engineering structures.

Three-Point Perspective:

In three-point perspective, the object is placed so that none of its principal edges are parallel to the Picture Plane, therefore, each of the three sets of parallel edges will have a separate Vanishing Point. The Picture Plane is assumed approximately perpendicular to the centreline of the cone of rays.

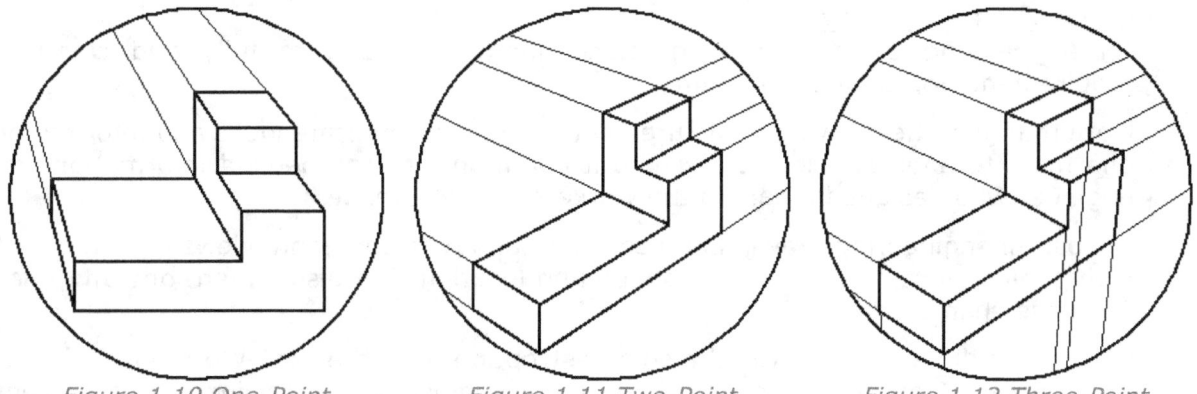

| *Figure 1.10 One-Point* | *Figure 1.11 Two-Point* | *Figure 1.12 Three-Point* |

Schematic Diagram:

A schematic diagram represents the elements of a system using graphical symbols rather than realistic and detailed drawings. A schematic usually omits all details that are not relevant to the information the schematic is intended to convey and may add unrealistic elements that aid comprehension. For example, a suburban bus map intended for passengers may represent a bus stop with a dot; the dot does not resemble the actual station at all but gives the viewer information without unnecessary visual clutter. A schematic diagram of a chemical process uses symbols to represent the vessels, piping, valves, pumps, and other equipment of the system, emphasizing their interconnection paths and suppressing physical details. In an electronic circuit diagram, the layout of the symbols may not resemble the layout in the physical circuit. In the schematic diagram, the symbolic elements are arranged to be more easily interpreted by the viewer.

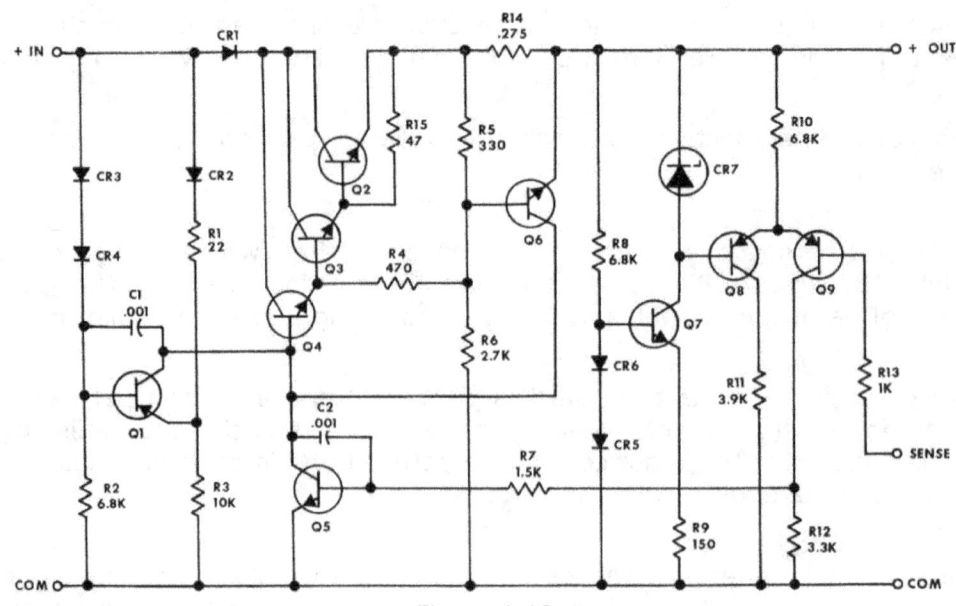

Figure 1.13

Critical Information:

An engineering drawing, is a graphical language, used to fully and clearly define requirements for engineering items.

More than just the drawing of pictures, it used to communicate ideas and information from one mind to another. Most especially, it communicates all needed information from the engineer who designed a part to the workers who will make it.

Almost all engineering drawings (except perhaps reference-only views or initial sketches) communicate not only geometry (shape and location) but also dimensions and tolerances for those characteristics.

Coordinate dimensioning was the sole best option until the post-World War II. The Need for Precise Communications developed the engineering drawings to the geometric dimensioning and tolerancing (GD&T), which departs from the limitations of coordinate dimensioning.

Drawings convey the following critical information:
- Geometry – the shape of the object; represented as views. The basis for much engineering drawing is orthographic representation (projection).
- Dimensions – the size of the object is captured in accepted units.
- Tolerances – the allowable variations from the nominal size for each dimension.
- Material – represents what the item is made of.
- Surface finish – specifies the surface quality of the item.
- Notes - written instructions which cannot be covered in a drawing view, e.g. the plate must be "hot dip galvanised" or "prepared surfaces applied with a lubricant".

Skill Practice Exercises:

Skill Practice Exercise MEM09002-RQ-0101:
Identify the following drawing types:

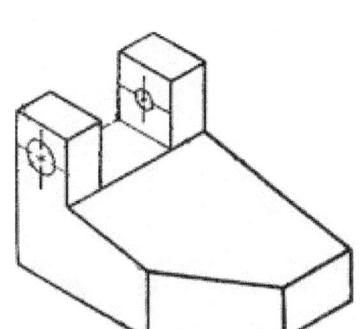

A. _____

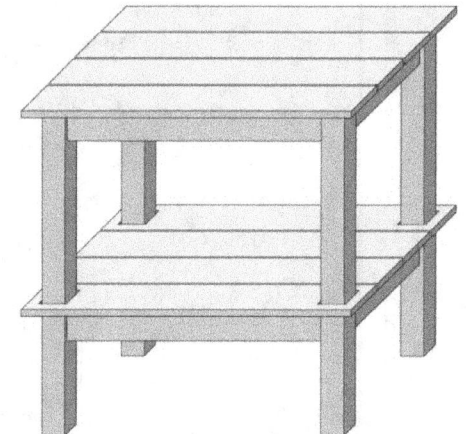

B. _____

C. _____

D. _____

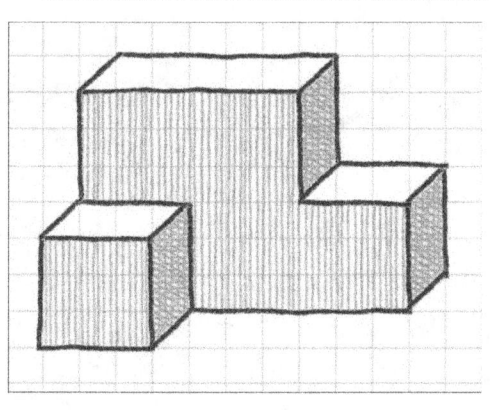

E. _____

F. _____

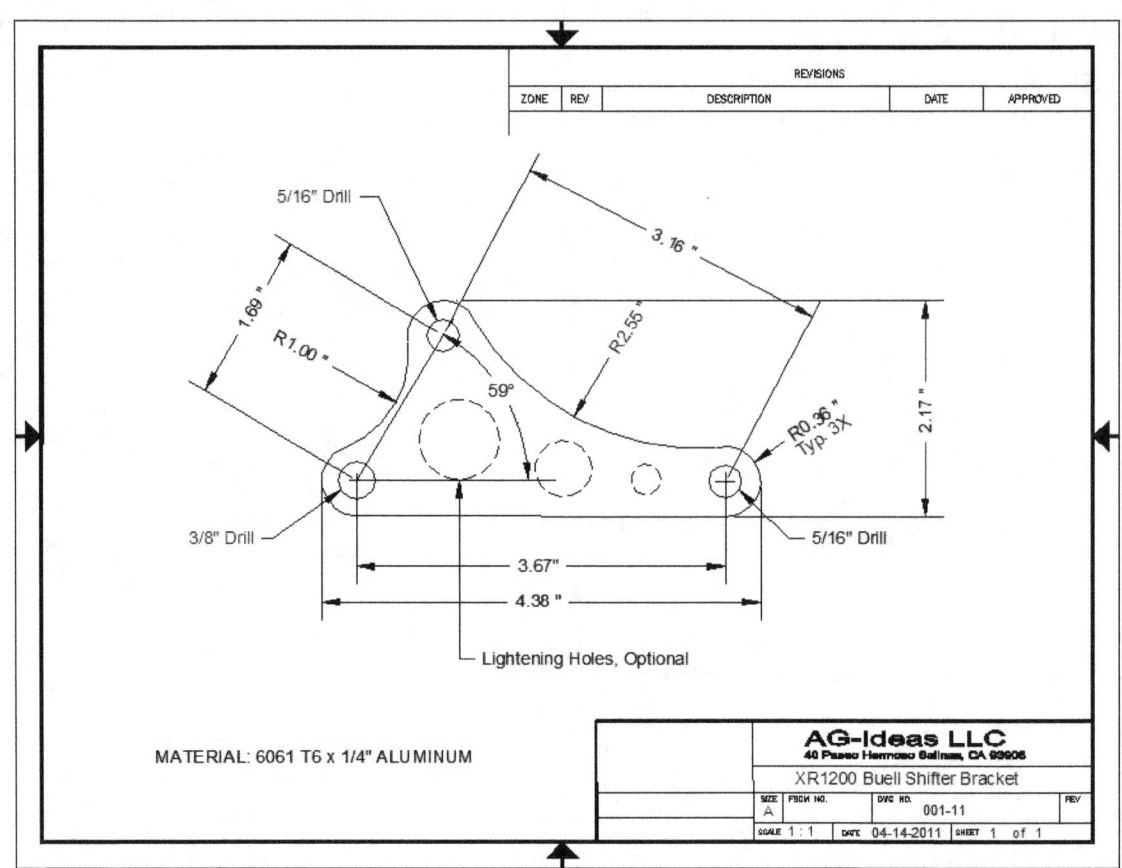

G. _____

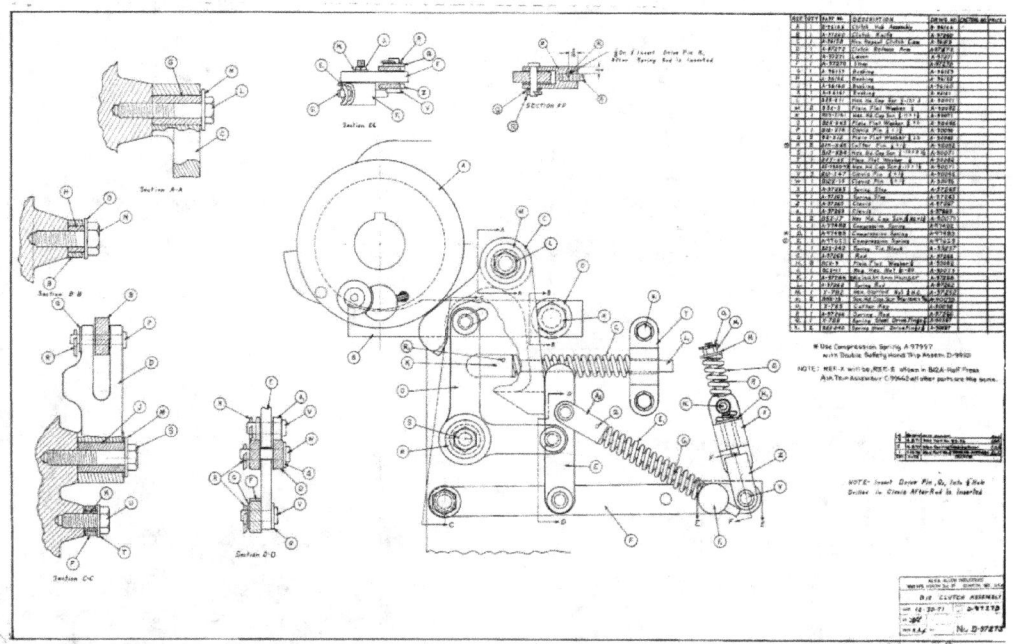

H. _____

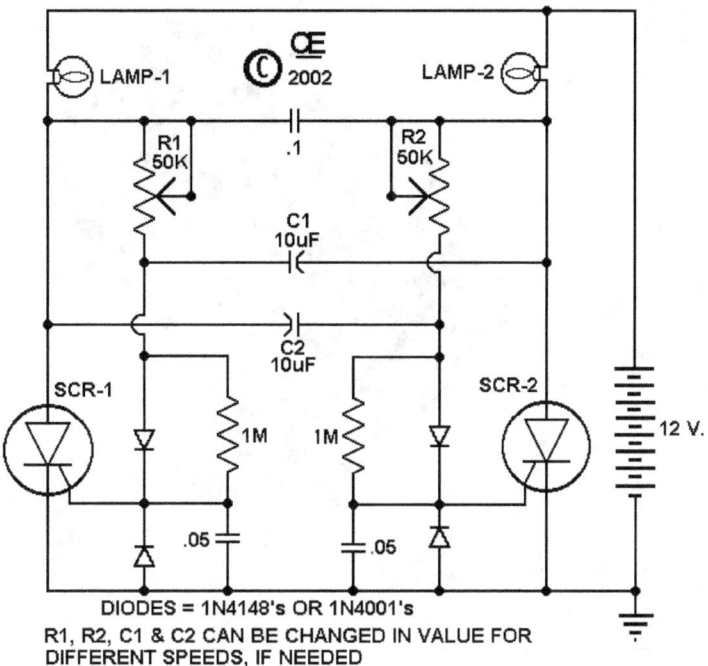

J. _____

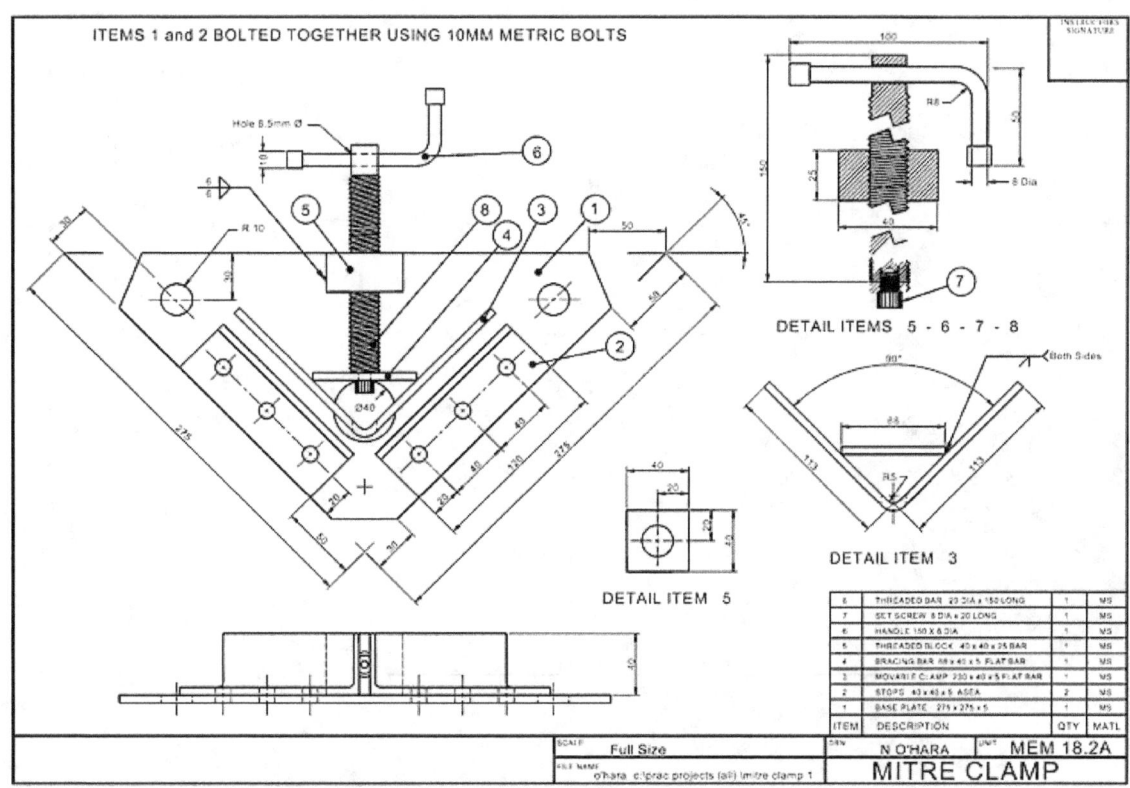

K. _____

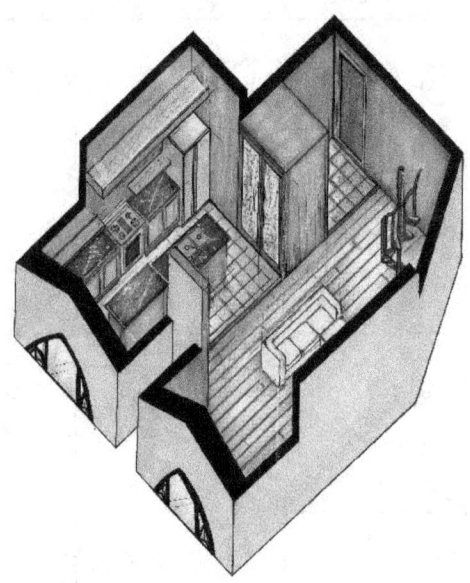

L. _____

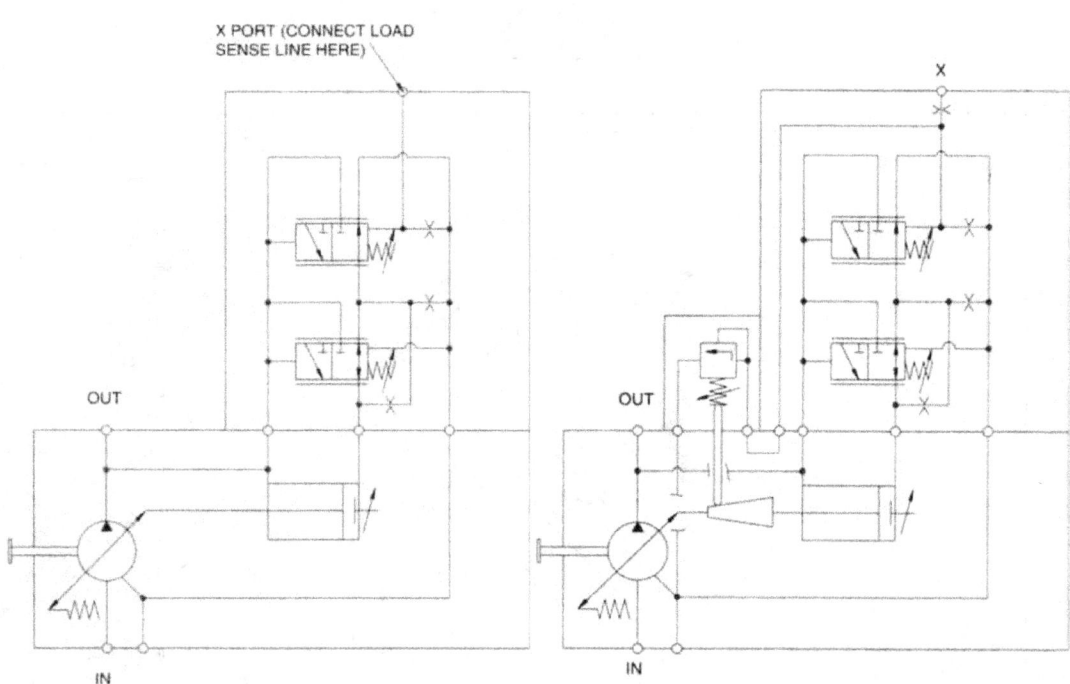

M. _____

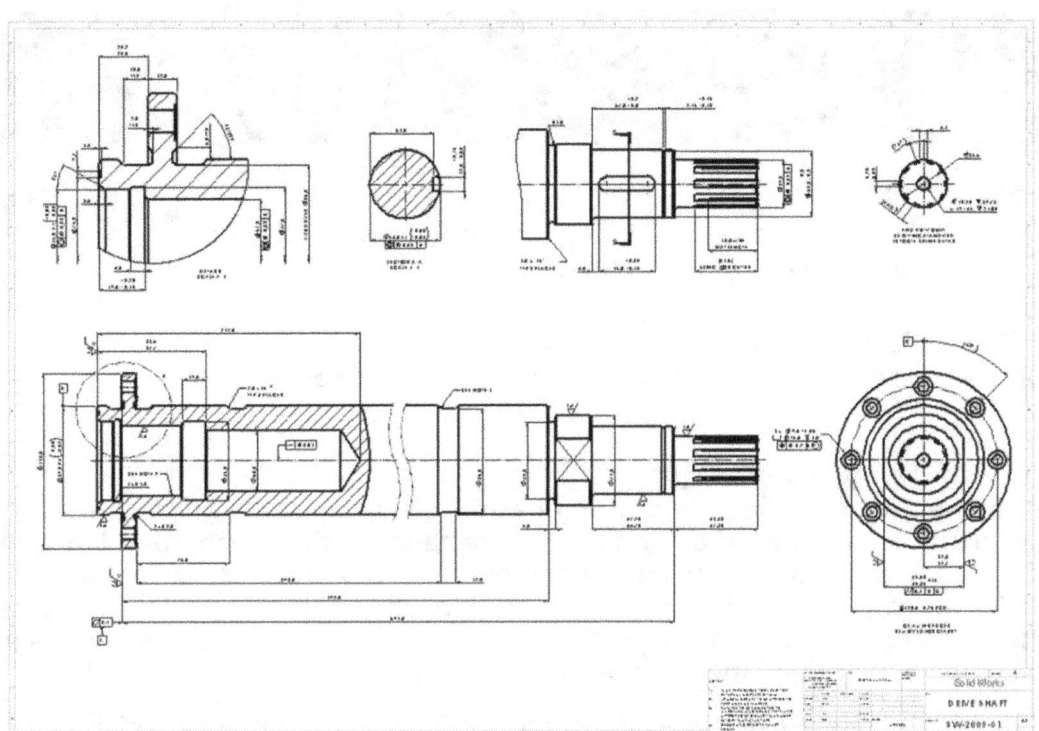

N. _____

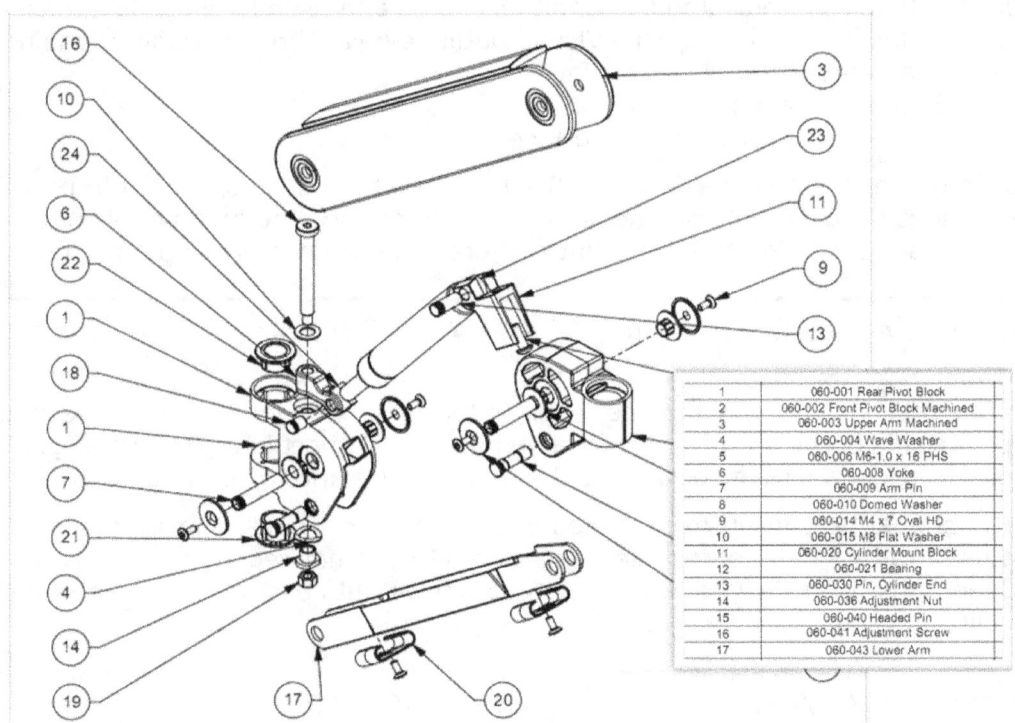

1	060-001 Rear Pivot Block
2	060-002 Front Pivot Block Machined
3	060-003 Upper Arm Machined
4	060-004 Wave Washer
5	060-006 M6-1.0 x 16 PHS
6	060-008 Yoke
7	060-009 Arm Pin
8	060-010 Domed Washer
9	060-014 M4 x 7 Oval HD
10	060-015 M8 Flat Washer
11	060-020 Cylinder Mount Block
12	060-021 Bearing
13	060-030 Pin, Cylinder End
14	060-036 Adjustment Nut
15	060-040 Headed Pin
16	060-041 Adjustment Screw
17	060-043 Lower Arm

O. _____

Name: _____

Topic 2 – Line Styles:

Required Skills:
- Types of lines used in engineering drawings.
- Precedence of lines on a drawing.

Required Knowledge:
- Line construction, widths and/or thicknesses.

Line Styles & Conventions:
Each line on an engineering drawing has a definite meaning and is drawn to a particular construction. The use of different line styles and widths is important in technical drawing as they are used to indicate details and features in a drawing. Line styles make drawings easier to read: for example, solid lines used to show the object will stand out from broken lines used to show hidden information. The correct usage of line styles is essential whether using manual drafting methods or CAD.

Line weight is the thickness of the lines and corresponds to the sheet size being used. AS 1100 incorporates a detailed list of line styles for use in different fields of design including architecture and engineering.

Visible Outlines:
Visible outlines are indicated by continuous lines and used to show all edges viewed by the eye when looking at the object. Visible outlines have three possible meanings:
- The intersection of two surfaces.
- Edge view of a surface.
- Contour view of a curved surface.

Visible outlines in CAD have 0.5 or 0.7 lineweights and when drawn in pencil, a heavy or thick line is used. Visible outlines are the predominant line on an engineering drawing and ALWAYS appear in front of other lines where several lines are superimposed.

_____

Paper Size A4/A3/A2 – 0.5 mm   *A1/A0 – 0.7 mm*

Hidden Outlines:
Hidden lines indicate any surface, edge or feature that cannot be seen because it is on the opposite side of the view from or located inside the object being drawn, e.g. a hole. Hidden lines must join a visible (or another hidden) outline and form "T" and "L" intersections.

Views should be chosen to show features with visible lines where possible. Hidden lines are used to make the object clearer and omitted if not required or can confuse the view. Hidden outlines in CAD have 0.25 or 0.35 lineweights and when drawn in pencil a light or thin line is used.

____ ____ ____ ____ ____ ____ ____ ____ ____ ____ ___

Paper Size A4/A3/A2 – 0.25 mm   *A1/A0 – 0.35 mm*

Centrelines:
Centrelines are used to indicate the axes of symmetrical features, circles, holes and paths of motion. The large dashes should cross at the intersection of centrelines and extend uniformly outside the feature for which they are drawn. Centrelines must always start and end with long dashes except for small holes where they can be continuous.

Centrelines in CAD have 0.25 or 0.35 lineweights and when drawn in pencil a light or thin line is used.

Paper Size A4/A3/A2 – 0.25 mm   *A1/A0 – 0.35 mm*

Dimension Lines:

A dimension line is used to define the measurement of a part feature. Dimension lines consist of a solid line with arrows at both ends and a dimension in the centre.

Dimension in CAD have 0.25 or 0.35 lineweights and when drawn in pencil a light or thin line is used.

Paper Size A4/A3/A2 – 0.25 mm   *A1/A0 – 0.35 mm*

Extension/Projection Lines:

An extension or projection line is used to visually connect the ends of a dimension line to the relevant feature on the part. Extension lines are drawn perpendicular to the dimension line.

Extension/Projection lines in CAD have 0.25 or 0.35 lineweights and when drawn in pencil a light or thin line is used.

Paper Size A4/A3/A2 – 0.25 mm   *A1/A0 – 0.35 mm*

Break Lines:

A break line indicates that a portion of the item is not shown on the drawing and is necessary for reasons of space or drawing clarity. Break lines can be straight or curved

Break lines in CAD have 0.25 or 0.35 lineweights and when drawn in pencil a light or thin line is used.

Paper Size A4/A3/A2 – 0.25 mm   *A1/A0 – 0.35 mm*

Cutting Plane Line:

The cutting plane line indicates the location of a view from where the Sectional View is taken. The arrowheads indicate the direction in which the cutaway object is viewed.

Cutting plane lines in CAD have a 0.5 line indicating the cutting plane and an arrow on a 0.25 or 0.35 lineweights and when drawn in pencil a thick and thin lines are used.

Paper Size A4/A3/A2 – 0.25 mm   *A1/A0 – 0.35 mm*

Phantom Lines:

Phantom lines are used most frequently to indicate an alternate position of a moving part. The part in one position is drawn in full lines, while in the alternate position it is drawn in phantom lines. Phantom lines are also used to indicate a break when the nature of the object makes the use of the conventional type of break unfeasible

Phantom lines in CAD have 0.25 or 0.35 lineweights and when drawn in pencil a light or thin line is used.

Paper Size A4/A3/A2 – 0.25 mm   *A1/A0 – 0.35 mm*

Existing or Adjacent Parts.

Existing or adjacent parts represent any structure or part immediately in the vicinity of the object. Existing or adjacent part lines in CAD have 0.25 or 0.35 lineweights and when drawn in pencil a light or thin line is used.

Paper Size A4/A3/A2 – 0.25 mm   *A1/A0 – 0.35 mm*

Typical Example of Line Styles:

The drawing shows the outline of a Plate with a series of drilled holes. For clarity, the dimensions are shown as smaller text while the larger text indicates the type of line.

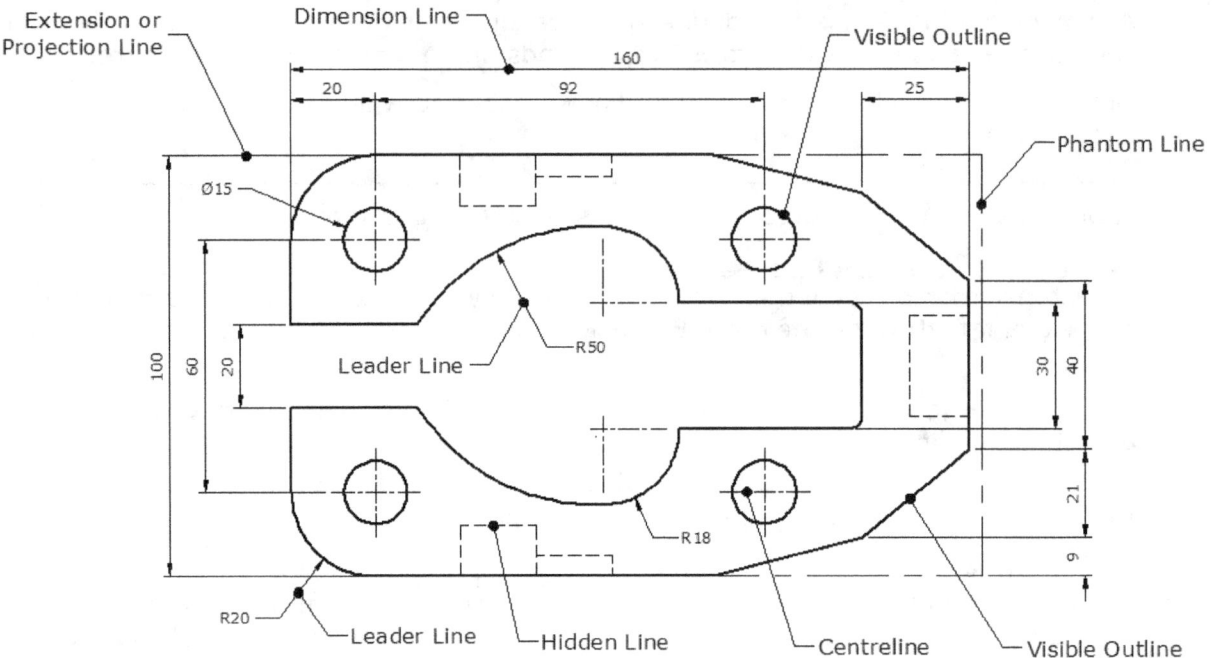

Precedence of Lines:

On any drawing two or more line styles are going to coincide. Hidden portions of the object may project to coincide with visible portions while centrelines may occur where there is a visible or hidden outline of some feature.

Since the physical features of the object must be represented, full and dashed lines take precedence over all other lines. AS visible outlines are more prominent then dashed lines, they take precedence; visible outlines can cover a hidden outline line but a hidden outline cannot cover a visible outline.

When centrelines and cutting plane lines coincide, the one that is more important takes precedence over the other. Break lines should be placed so they do not spoil the readability of the drawing. Dimension and projection lines must always be placed so as not to coincide with other lines on the drawing.

The precedence of lines are:
- Visible outline
- Hidden outline
- Centreline or Cutting plane line
- Break line
- Dimension & Projection line
- Cross hatching

Correct   **Wrong**

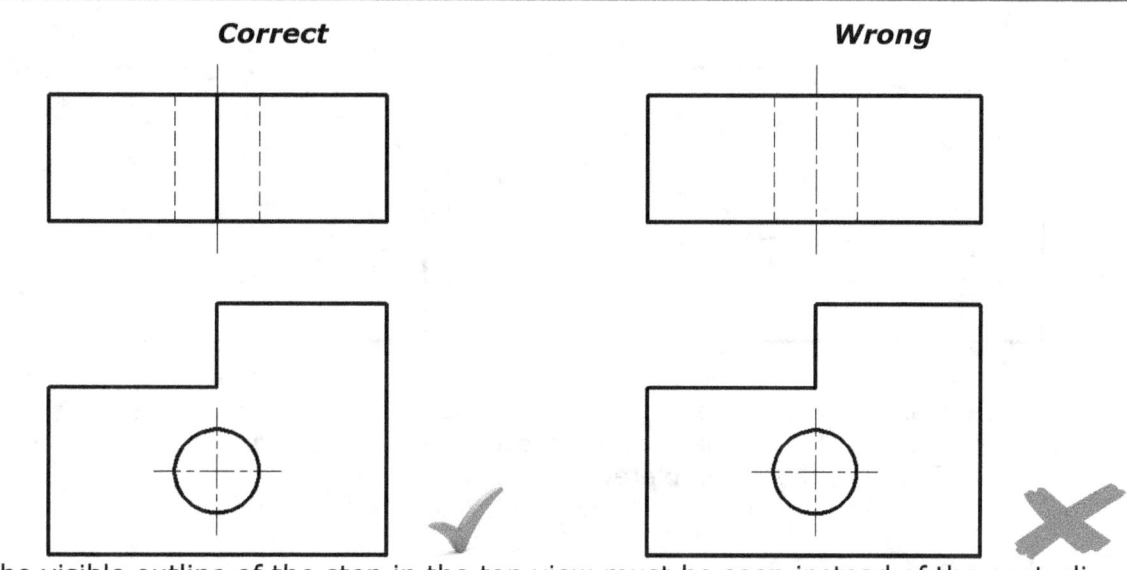

The visible outline of the step in the top view must be seen instead of the centreline.

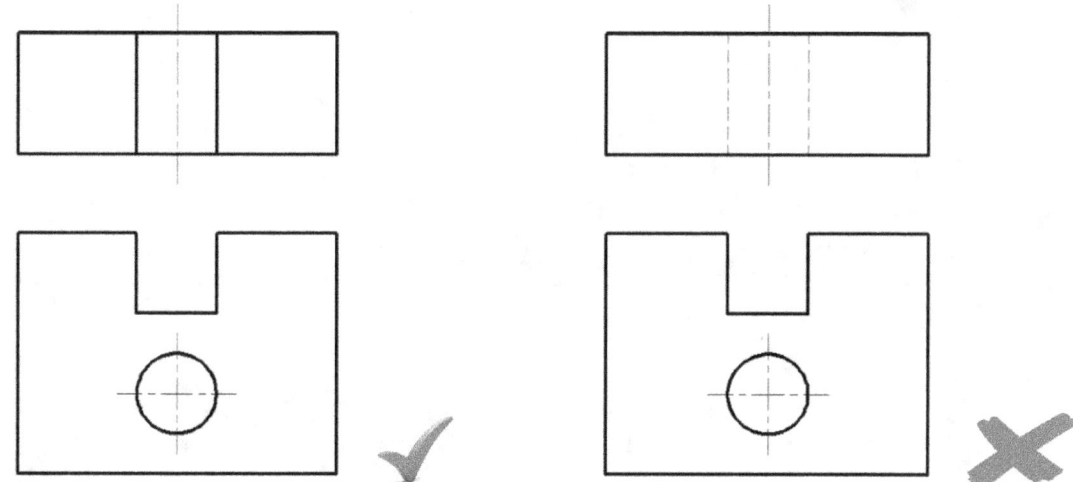

The visible outlines of the cut out in the top view take precedence over the hidden outlines of the drilled hole.

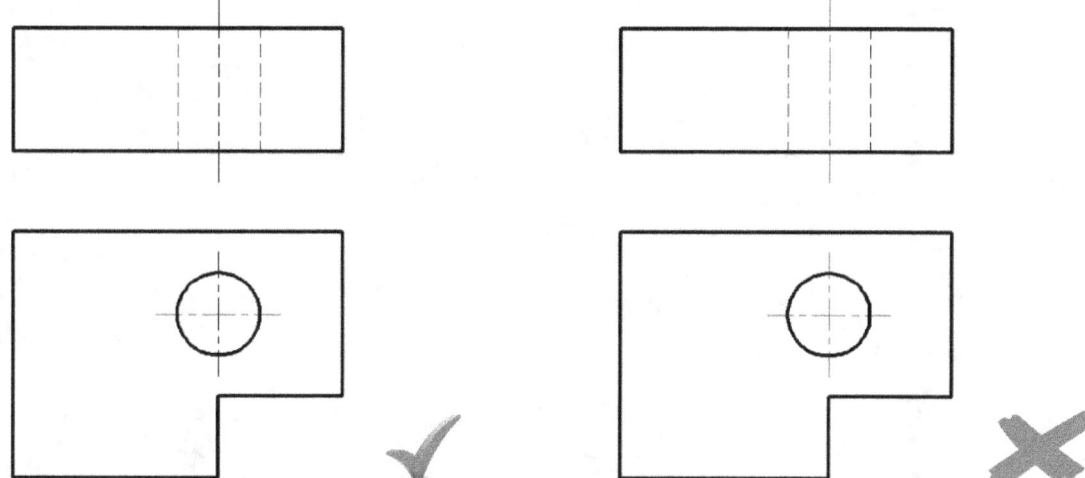

Although the centreline is not fully shown, the extensions can be clearly seen in the top view while the hidden lines indicate the drilled hole and the hidden step on the bottom edge. In the wrong example it appears as if there is no hidden detail of the step below the drilled hole.

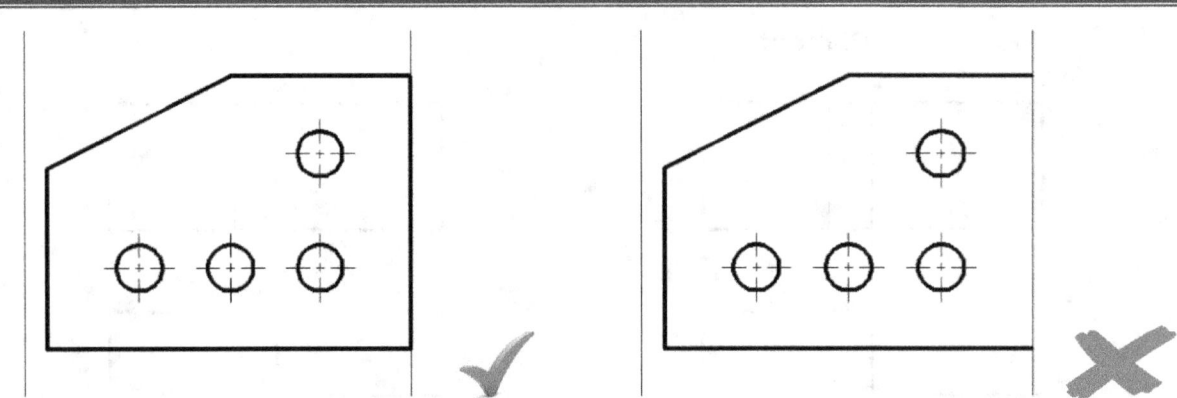

The above example shows a new gusset plate to be attached to an existing structure. In the correct example the full shape of the gusset plate can be seen while the wrong example appears as if the detail is incomplete.

Skill Practice Exercises:

Skill Practice Exercise MEM09002-RQ-0201:
Referring to the following images, name the indicated line style in the space provided.

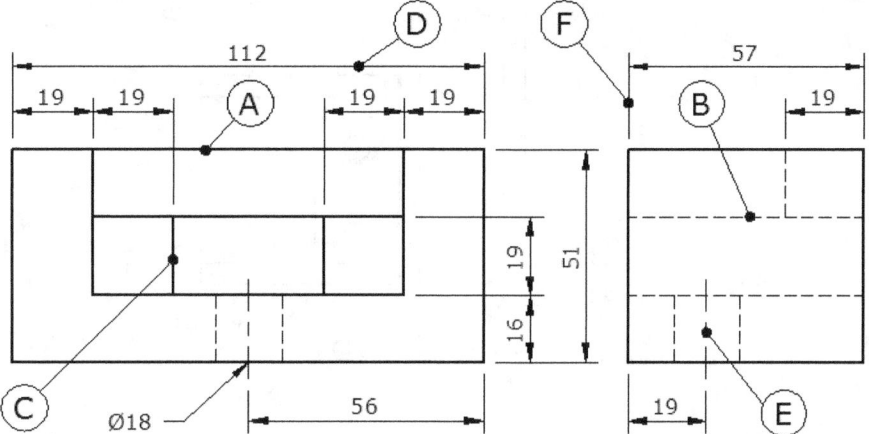

A. _____   B. _____

C. _____   D. _____

E. _____   F. _____

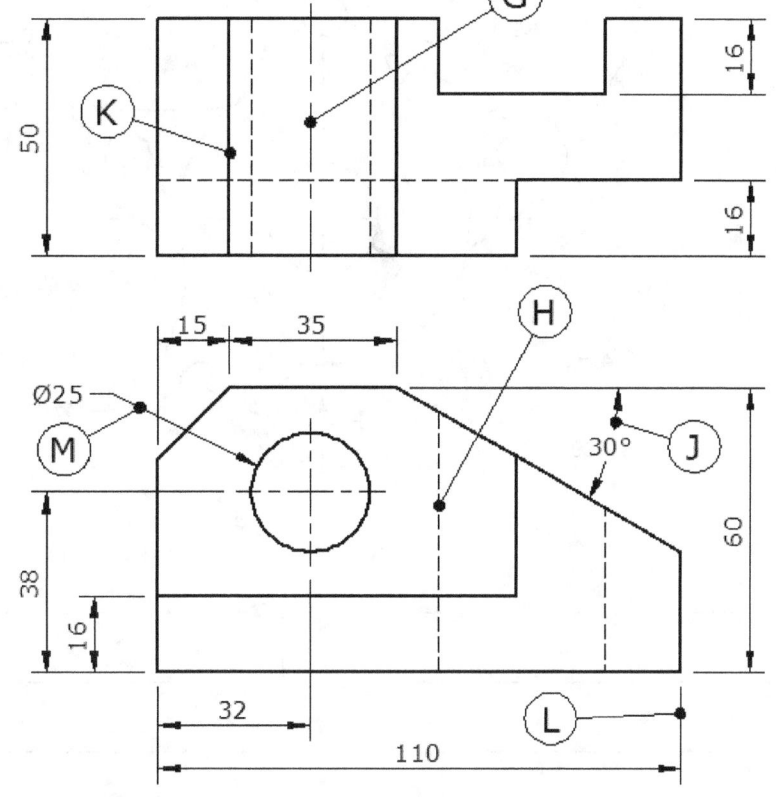

G. _____   H. _____

J. _____   K. _____

L. _____   M. _____

Skill Practice Exercise MEM09002-RQ-0202:

Referring to the following images, name the indicated line style in the space provided.

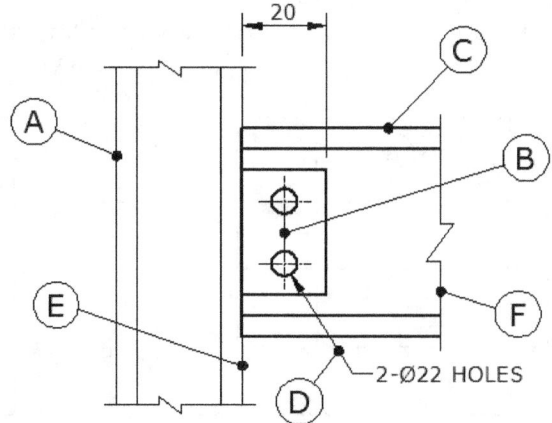

A. _____   B. _____

C. _____   D. _____

E. _____   F. _____

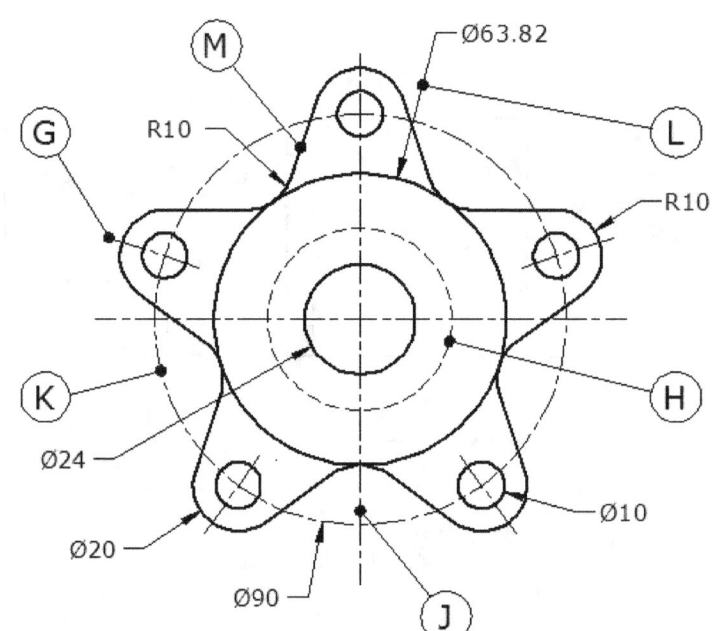

G. _____   H. _____

J. _____   K. _____

L. _____   M. _____

Name: _____

Topic 3 – Reading Drawings:

Required Skills:
- Checking the drawing against job requirements/related equipment in accordance with standard operating procedures.
- Confirming the drawing version as being current in accordance with standard operating procedures.
- Reading, interpreting information on the drawing.
- Checking and clarifying task related information.

Required Knowledge:
- Application of AS1100 in accordance with standard operating procedures.
- Relationship between the views contained in the drawing.
- Understanding of the instructions contained in the drawing.
- The materials from which the object(s) are made.

Reading Engineering Drawings:
Engineering drawings are typically used as visual tools in the creation of bridges, towers, airplanes, ships and boats, motor vehicles, gearboxes, braking systems, conveyors, residential industrial and commercial buildings etc.; the range of applications is nearly endless. Anything that is to be fabricated, manufactured, or constructed MUST have drawings prepared for planning, estimating, costing, material ordering before any actual the workshop is provided with the drawings for work to commence.

While these drawings can be quite straightforward to individuals who are skilled in the field of engineering or architecture, they can be quite difficult to interpret for laypeople. Knowing how to read engineering drawings will help to provide a better idea of the drawings or plans. The key to interpreting engineering drawings is to understand the purpose of a specific drawing and the relationship of that drawing to the overall set of engineering drawings and specifications prepared for a project.

Reading the orthographic language is a mental process as the drawing is not read aloud unless discussing the drawing with other workers. To describe even a simple object with words is almost impossible.

"A picture is worth a thousand words."

Reading proficiency develops with experience, as similar conditions and shapes occur so often that a person in the field gradually acquires a background of knowledge that enables them to visualize readily the shapes shown. Experienced readers read quickly because they can draw upon their knowledge and recognize familiar shapes and combinations without hesitation. However, reading a drawing should always be done carefully and deliberately, as a whole drawing cannot be read immediately any more than a whole page of a printed book, magazine, or newspaper.

Prerequisites and Definitions:
Before attempting to read a drawing, familiarize the reader MUST be familiar with the principles of orthographic projection. Keep constantly in mind the arrangement of views and their projection, the space measurements of height, width, and depth, what each line represents, etc.

Visualization is the medium through which the shape information on a drawing is translated to give the reader an understanding of the object represented. The ability to visualize is often thought to be a "gift" that some people possess and others do not; this however, is not true. Any person of reasonable intelligence has a visual memory, as can be seen from their ability to recall and describe scenes at home, actions at sporting events.

The ability to visualize a shape shown on a drawing is almost completely governed by a person's knowledge on the principles of orthographic projection. The common adage that "the best way to learn to read a drawing is to learn how to make one" is quite correct, because in learning to make a drawing one is forced to study and apply the principles of orthographic projection. Reading a drawing can be defined as the process of recognizing and applying the principles of orthographic projection to interpret the shape of an object from the orthographic views.

Method of Reading:

A drawing is read by visualizing units or details one at a time from the orthographic projection and mentally orienting and combining these details to interpret the whole object finally. The form taken in this visualization, however, may not be the same for all readers or for all drawings. Reading is primarily a reversal of the process of making drawings; and since drawings are usually first made from a picture of the object, the beginner often attempts to carry the reversal too completely back to the pictorial. The result is that the orthographic views of an object like those covered in Topic 4 – Orthogonal Projection:

To most, it is a mental impossibility (and surely unnecessary) to translate more than the simplest set of orthographic views into a complete pictorial form that can be pictured in its entirety. The reader goes through a routine pattern of procedure. Much of this is done subconsciously; for example, consider the object in Figure 3. 1. A visible circle is seen in the top view. Memory of previous projection experiences indicates that this must be a hole or the end of a cylinder. The eyes rapidly shift back and forth from the top view to the front view, aligning features of the same size ("in projection"), with the mind assuming the several possibilities and finally accepting the fact that, because of the dashed lines and their extent in the front view, the circle represents a hole that extends through the prism.

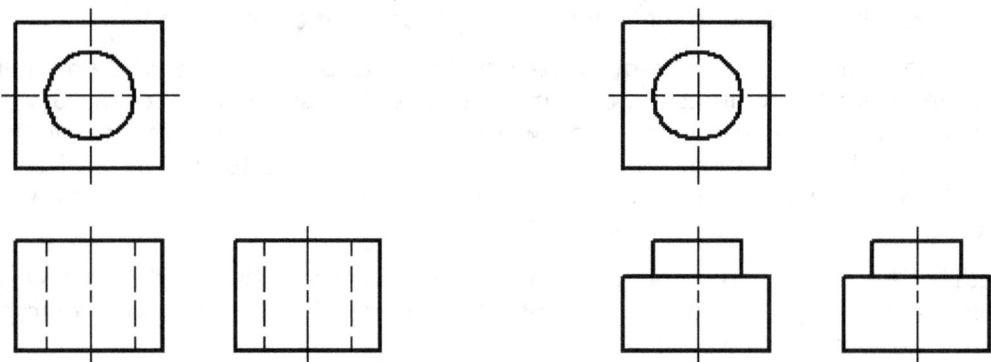

Figure 3. 1   Figure 3. 2

Following a similar pattern of analysis, the reader will find that Figure 3. 2represents a rectangular prism surmounted by a cylinder; this thinking is again done so rapidly, that the reader is scarcely aware of the steps and processes involved.

The following is a suggested technique for reading a drawing; but how does the beginner develop the ability?

First:
The reader must have a proficient working knowledge of the principles of orthographic projection.

Second:
The reader must acquire a complete understanding of the principles behind the meaning of lines, areas, etc., and the mental process involved in interpreting them, as these principles are applied in reading.

There is little additional learning required. Careful study of all these items plus practice will develop the ability and confidence needed.

Procedure for Reading:

The actual steps in reading are not always identical because of the wide variety of drawings. Nevertheless, the following outline gives the basic procedure and will serve as a guide:

First:

The reader should orientate themself with the views given.

Second:

The reader should obtain a general idea of the over-all shape of the object. Think of each view as the object itself, by visualizing being in front, above, and at the side as is done in making the views. Study the dominant features and their relationship to one another.

Third:

Start reading the simpler individual features, beginning with the most dominant and progressing to the subordinate. Look for familiar shapes or conditions that your memory retains from previous experience. Read all views of these familiar features to note the extent of holes, thickness of ribs and lugs, etc.

Fourth:

Read the unfamiliar or complicated features. Remember that every point, line, surface, and solid appears in every view and that the projection of every detail in the given views must be located to learn the shape.

Fifth:

As the reading proceeds, note the relationship between the various portions or elements of the object. Such items as the number and spacing of holes, placement of ribs, tangency of surfaces and the proportions of hubs etc., should be noted and remembered.

Sixth:

Reread any detail or relationship that is not clear at the first reading.

Important Information to Understand:

There is a lot of important information stored on a drawing sheet and is contained in the detail and assembly views, and the Title, Revision and Materials/Parts/Cutting Lists.

Title Block:

When preparing a set of drawings for planning, construction, or manufacture purposes, one thing that is required on every single drawing is a Title Block. This is simply a set of information that clearly identifies common information and includes all the information which enables the drawing to be interpreted, identified and archived. Title Blocks can run horizontally across the bottom of the page or vertically situated on the right side of the page.

The minimum important information to be read from a Title Block are drawing title, scale, date, drawing number, revision number and checker:

Drawing Title:
The drawing title must fully describe the drawing, e.g. RT425 Radar Module Foundation Plinth, not Foundation Plinth or Radar Foundation.

Drawing Number:
The drawing number is a number given to a drawing for identification and reference. Every drawing should have a unique drawing number. The number system is different from one company to another; for instance, in one company the drawings could start at A00 then the next would be A01, A02, A03 etc. The 'A' at the start of the number in this case would represent the 'A' in the word 'Architectural' to signify an Architectural drawing. This can differentiate the Architectural drawings from other drawings from consultants such as Structural Engineers that often have drawings such as S01, S02, S03 etc. Another company the numbering system could show FGT-10P-082 where FGT refers to the project, 10P to the type of drawing, and 082 the drawing in the sequence starting from 001.

Scale:

The shows at what scale the drawing has been plotted/printed at. The scale would normally read 1:1, 1:20, 1:100, 5:1

NEVER scale from a printed drawing as it MAY be printed to a smaller or larger scale; even printing at 99% will give an incorrect measurement. All measurements used for construction should be obtained by the written dimensions on the drawings for accuracy. If in doubt about a measurement, ask the person who drew the drawings.

At times, a drawing may be produced on an A3 sheet for printing on an A4 sheet; the scale may read NTS meaning "NOT TO SCALE". Schematic diagrams are normally NTS.

Revision Number:

Any changes made to a drawing after it has been approved, printed and circulated must be recorded on a drawing, whether it is a design change or even a mistake. A description is placed in the Revision Block but the revision number is always placed in the Title Block. The first revision will be marked as A then the next would be B etc.; the first letter A may also refer to the original issue of the drawing with the first revision being shown as B.

Drawn:

It is normally a requirement for the person who produces the drawing has their initials shown in the Drawn box as part of the office protocol. Depending on the drafting discipline, most companies only require initials to be shown but some may require the officer's name, e.g. J.F.K. or JOHN F. KENNEDY.

Checked/Examined/Approved:

Once a drawing is submitted to a council or building surveyor, this

Date:

The date normally signifies when the drawing was completed by the draftsperson. Other dates should also refer to when the drawing was checked, examined, approved etc., by other supervisors, and engineers or architects. The date should never be changed on the original drawing as it can be used for legal purposes if the need arises.

Revision Block:

During the life of a product, components and assemblies that are used for many years, they may be revised several times to improve performance or reduce cost. After a drawing change request is made, the drawing is modified. Any change to a drawing after the previous release, must be recorded in the Revision Block.

Revision Blocks are simplified to show the revision number, a description of the change, the date and either the drafting officer responsible for the change or their checker.

Revision Number:

The revision number must be the same as shown in the revision part of the Title Block. A, B, C etc.

Description:

The description briefly describes the change e.g. "Ø20 changed to Ø25" or "Material changed from M.S. to H.T.S."

Date:

The date normally signifies when the drawing was completed by the draftsperson or checked/approved by a supervisor for release.

Drawn/Checked:

The initials of the drafting officer or the checker/approver must be placed in the box.

Materials/Parts/Cutting Lists:

A Materials or Parts List, also known as a bill of materials (BOM) and is a tabular list of the items used to make an assembly. Parts list is usually combined with the assembly drawing, but it is a separate and individual document and can be and provides a complete list of all parts needed to build the complete project. Another type of the lists is related to the fabrication industry and is called a Cutting list where the different sizes of metal plates and structural sections are listed.

The lists can be in the upper-right corner of the sheet, or above title block, or in a convenient location depending on the company's standard.

The information associated with the parts list generally includes item number, description, quantity and material but can be expanded to include catalogue number, drawing number, specification number, length, and notes to name but a few.

Item Number:
Item numbers are based on the assembly structure, that is, the order in which parts are displayed in assembly, e.g. 1, 2, 3, 4, etc.

Description:
Description is usually a part name or a complete description of purchase part or stock specification, including size and dimensions, e.g. BOLT HEX HD M10x0.7x75, FLAT BAR 50x10,

Quantity:
The number of the part used on the assembly, e.g. 1, 4, 16 etc.

Material:
The material of the component is normally abbreviated, e.g. MS (Mild Steel), ALAL (Aluminium Alloy), SS (Stainless Steel).

Catalogue Number:
A catalogue number is a special number a company assigns to components or parts for identification purposes, e.g. I/N: 2310582 (Bunnings), 220 E-2SR1TN9 (SKF Bearings)

Drawing Number:
Many components in a drawing may have been designed and detailed by the company and are available within the plan document centre. Those drawings have their own drawing numbers and can be identified in the appropriate column.

Specification Number:
All materials are produced to specific specifications. The specification number can be placed in the specification column to ensure the correct grade of material is procured, e.g. AS/NZS 1163:2016, AS 1830-2007, AS/NZS 1867-1997, ISO 5892:2013.

Length:
In a cutting list the width and thickness are shown in the description column while the length is often placed in its own column, e.g. 455, 1200, 2750.

Notes:
The notes column is used to describe any special features or treatments to the part, e.g. ULLAGE ITEM 4, GREASE MACHINE SURFACES.

Interpreting Drawings:
It is one thing to be able to read a drawing, but that information learned from the drawing must be interpreted.

Interpreting a Drawing in Preparation for Manufacture:
It is not usually the prerogative of the designer to decide the details of the machining of a component, although it is often possible to foretell the sequence of some of the manufacturing processes involved; from knowing the manufacturing sequence the designer can identify the manufacturing datum face(s)1, and from this, the required machining dimensions. The datum faces will be those faces used to hold the component during manufacture.

It is common practice in various industries to produce stage drawings. The datum features that are used to produce components stage by stage may not be the same as the finished drawing datum features. For example, a hole may have been produced in the component to allow for the product to line up on a fixture to produce other features, for example, turbine blades slots. Once the slots have then been produced, the hole could be in a feature that is then removed before the component is fully completed. The planning of the manufacturing process based on the stage or final drawings is vital to the success of producing quality component parts. Many areas have to be considered before the manufacturing process begins. Consideration of the following is important:

- What is the required method of manufacture?
- Availability of resources, such as machines, tooling, personnel, equipment.
- How do we hold the component?
- Is fixing required?
- Is in-line measurement used?
- Gauging or measuring instruments?
- What are the best instruments to use?
- Have influences of measurement uncertainty been considered?
- Will training be required?

Once these questions have been answered the production process can begin. It is now important to think about how to monitor the process; our current capability is known, but can the processes consistently be controlled? It may be that as part of the manufacturing process statistical techniques may be used to assist in the interpreting.

Interpreting a Drawing in Preparation for Measurement:

The importance of interpreting the design requirements cannot be stressed too highly during preparation for measurement. Identifying the geometric characteristics of the component and the datum features that make up the co-ordinate system is critical to a successful measurement strategy. When making measurements, make use of datum features identified in drawings, technical documents, or computer aided design (CAD) models that relate directly to the component.

Datum features on a drawing are normally an important characteristic – a locating or positioning feature. A datum feature could be a face (a surface), a centre line (an axis), or a series of characteristics that collectively make up a datum system. The datum system may be easy to set up when using conventional measuring equipment, such as a surface table in conjunction with angle plates, dial gauges, height gauges and gauge blocks.

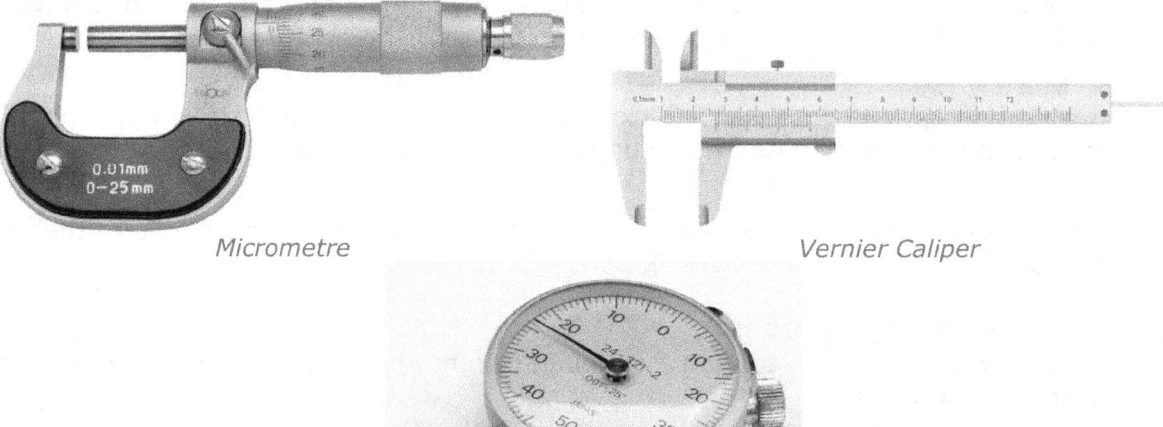

Micrometre   *Vernier Caliper*

Dial Gauge

Height Gauge | Gauge Blocks | Co-Ordinate Measuring Machine

Alternatively, the use of CAD data may be a requirement of the measurement process and, therefore, computerized measuring equipment such as the co-ordinate measuring machine (CMM), may need to be used.

When setting up a datum for measurement it is preferable to choose as datum features the surfaces that were used in the manufacturing process to hold the component; this choice relates the inspection results directly to the manufacturing process.

The features of any component can be defined in two ways, relative to a datum position or positions (absolute), or relative to one another (incremental). The co-ordinate system should be clearly defined whether on a physical drawing or CAD model.

From the drawing the measurement strategy must be determined for the geometric characteristics and the co-ordinate system. Account must be taken of environmental considerations such as temperature effects, equipment required and the associated uncertainties in relation to the stated specifications.

Skill Practice Exercises:

Skill Practice Exercise MEM09002-RQ-0301

Refer to drawing STPL-12H-36 and answer the following questions:

1. What grid zone is the Detail of the Oil Grove located?

   _____

2. What dimension and grid zone is the dimension drawn "Not to Scale"?

   _____

3. What type of section is shown in the right side view?

   _____

4. What are the overall dimensions of the Body?

   _____

5. How many surfaces are to be machined?

   _____

6. Name the method of projection used to produce the drawing.

   _____

7. What is the title of the drawing?

   _____

8. What size hole must be drilled for the M12x1 tapped hole?

   _____

9. What does PCD mean?

   _____

10. What is the general size for all unspecified fillets?

    _____

11. What is the centre-to-centre distance between the Ø22 holes?

    _____

12. What temperature and time must the Body be heated?

    _____

13. What is the general tolerance used on the drawing?

    _____

14. What is the diameter of the large vertical hole in the Body?

_____

15. What is the distance between the tolerance Ø165 and Ø24 holes?

_____

16. What change was done to the drawing for Issue C?

_____

17. Who checked the drawing?

_____

18. What drawing must be referred to obtain details of the Adaptor Plate?

_____

19. What scale is used to detail the Oil Groove drawn?

_____

20. When are holes to be tapped?

_____

21. What is the name of the company producing the drawing?

_____

22. What material is used to manufacture the Body?

_____

23. Determine the thickness of the base of the Body.

_____

24. What radius is used on the Oil Groove?

_____

25. What does U.N.O. mean?

_____

26. What type of tolerance dimensions are used throughout the drawing?

_____

Skill Practice Exercise MEM09002-RQ-0302
Refer to drawing HC145-58 and answer the following questions:

1. What is the title of the drawing?

   _____

2. How many pressure gauges are used in the system?

   _____

3. What is the latest revision number?

   _____

4. How many Relief Valves are used in the system?

   _____

5. How many 3 Position 4 Port directional Control Valves are used in the system?

   _____

6. Who drew the original drawing?

   _____

7. What is the drawing number?

   _____

8. Who checked the drawing and date was it checked?

   _____

9. How many Hydraulic Actuators are used in the system?

   _____

10. How many Check Valves are used throughout the circuit?

    _____

11. How many branch points are in the system indicated by dots (•)?

    _____

12. What change was done to the drawing in Revision C.?

    _____

13. How many 2 Position 4 Port directional Control Valves are used in the system?

    _____

14. What does NTS mean?

    _____

Name: _____

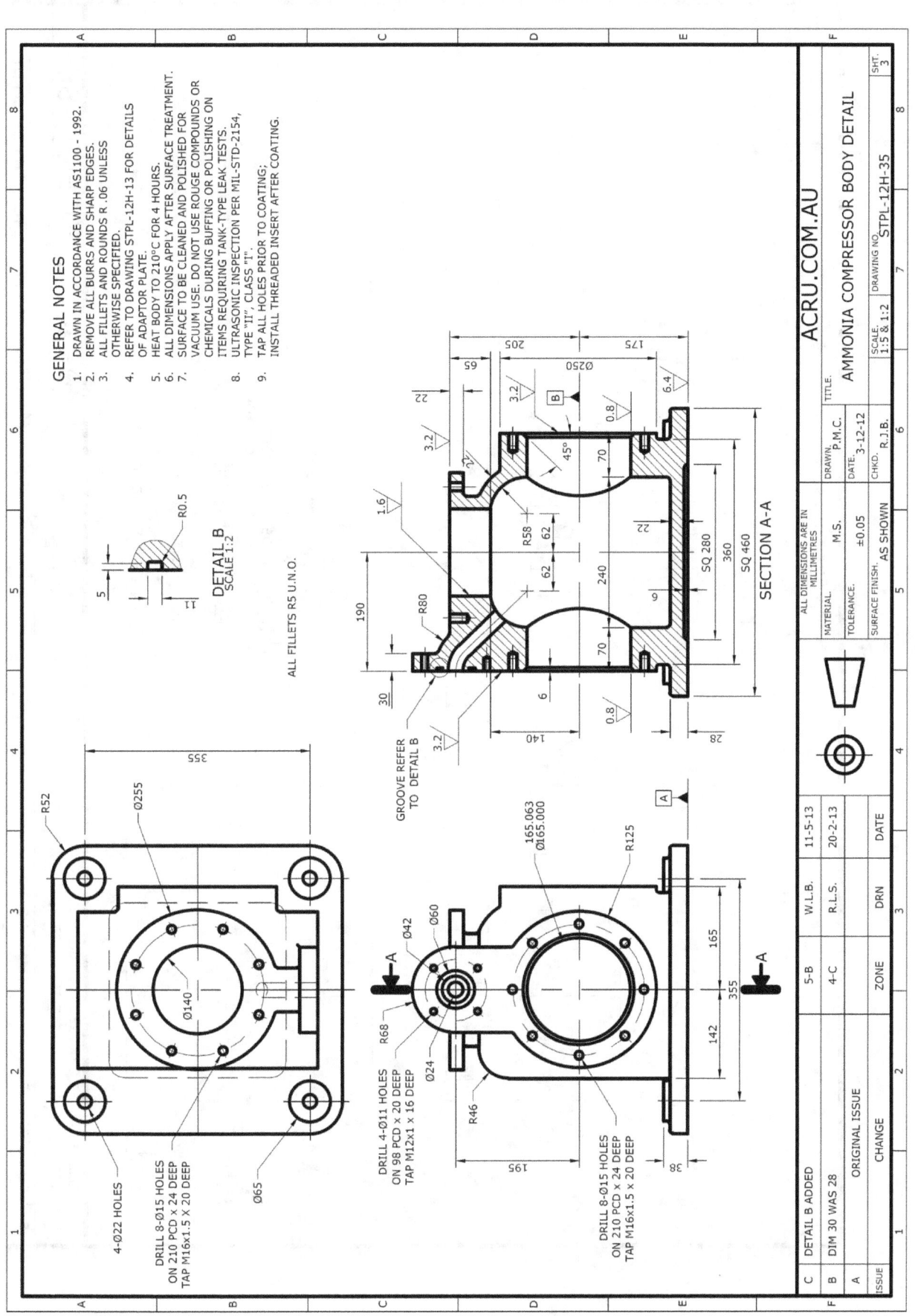

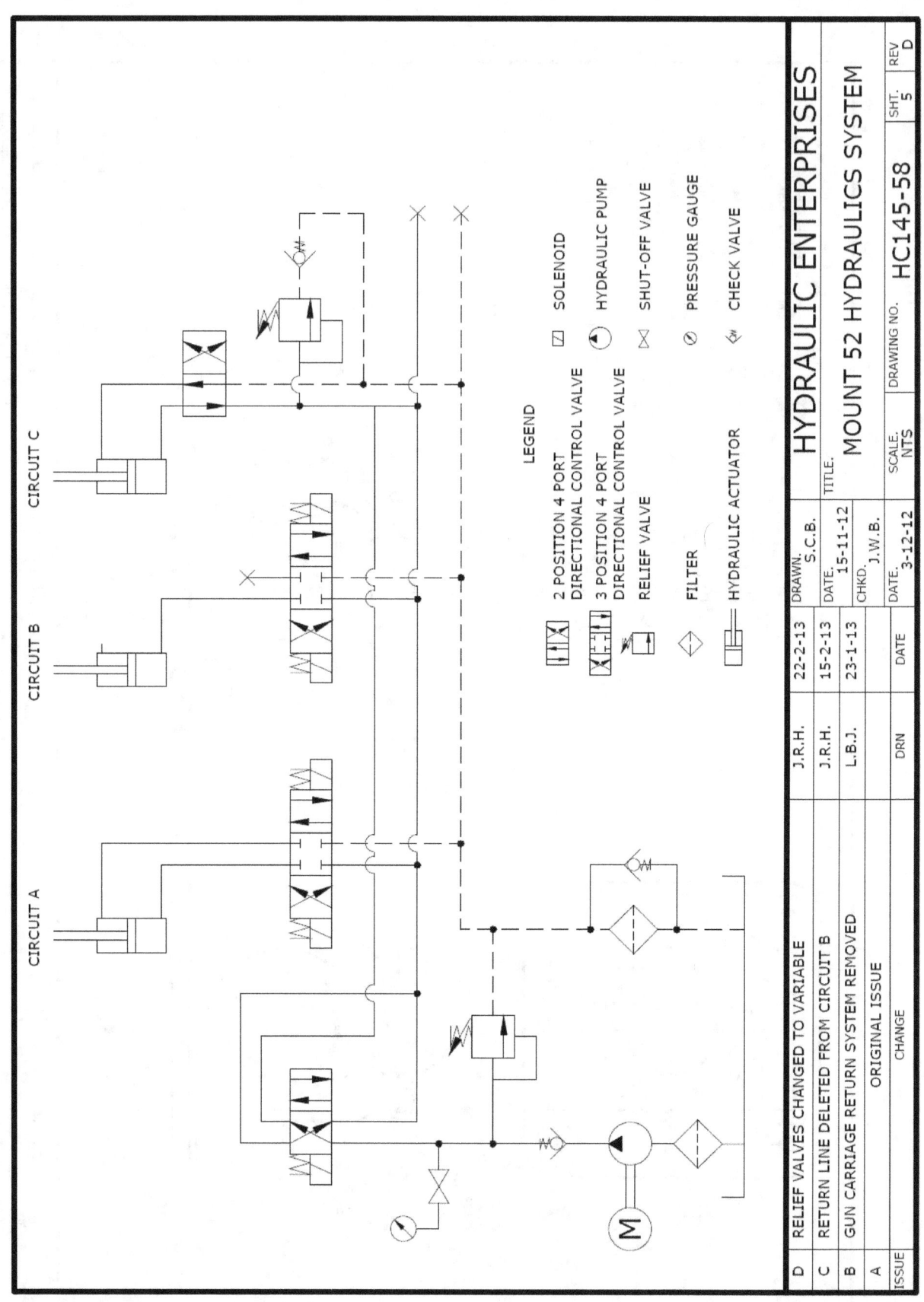

Topic 4 – Orthogonal Projection:

Required Skills:

- Name the methods used to produce engineering drawings.
- Identify projection symbols and distinguish the difference between First Angle and Third Angle Projection.
- Minimum number of views required to describe the component/assembly.

Required Knowledge:

- Application of AS1100.
- Relationship between the views contained in the drawing.
- Objects represented in the drawing.

NOTE:

This topic is intended to introduce the student to the types of projection, names, and positions of the views. The drawing of orthogonal views in engineering drawings is covered in units MEM09003 Prepare basic engineering drawing and MEM09005 Perform basic engineering detail drafting, and other drafting discipline related units.

Orthographic Projection:

Most drawings produced and used in industry are multiview drawings. Multiview drawings are used to provide accurate three-dimensional object information on two-dimensional media, a means of communicating all of the information necessary to transform an idea or concept into reality. The standards and conventions of multiview drawings have been developed over many years, which equip us with a universally understood method of communication.

Multiview drawings usually require several orthographic projections to define the shape of a three-dimensional object. Each orthographic view is a two-dimensional drawing showing only two of the three dimensions of the three-dimensional object. Consequently, no individual view contains sufficient information to completely define the shape of the three-dimensional object. All orthographic views must be looked at together to comprehend the shape of the three-dimensional object. The arrangement and relationship between the views are therefore particularly important in multiview drawings.

Orthographic projection is a system of drawing to represent 3-D objects by using multiple view drawings. The word "Ortho" is a Greek word that means right or true. In this system of projection, the 3-D object is projected perpendicularly onto a projection plane with parallel projectors as shown in Figure 4. 1.

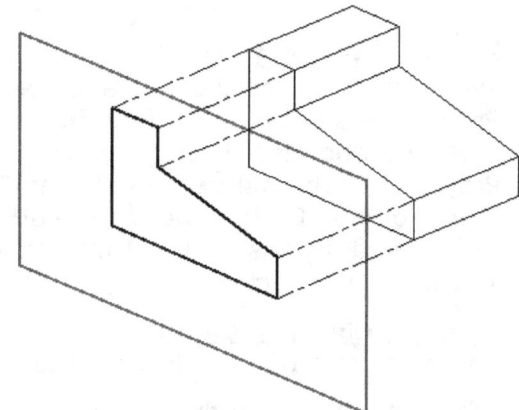

Figure 4. 1

Basic Views:

All objects can be projected in six orthogonal directions (Figure 4.2). The resulting views are called basic views. Orthogonal Projection can be thought of as a 3D object being placed inside a transparent box, and views projected orthogonally onto the six walls of the box.

The basic views are:
- Front View
- Top View
- Right & Left Views
- Bottom View
- Rear View

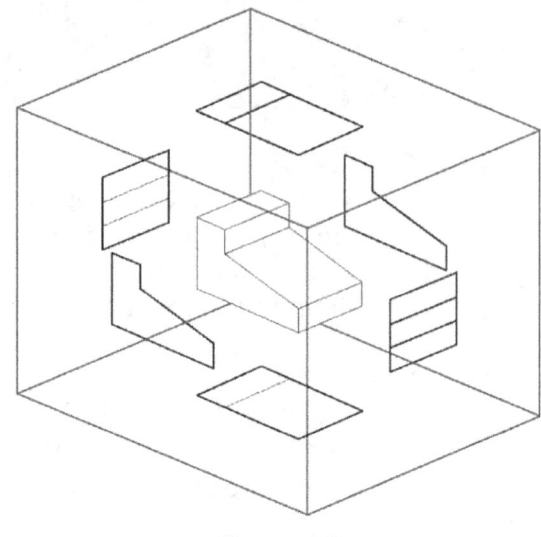

Figure 4.2

Developing the Box:

The transparent box may suit a "virtual Reality" environment; it cannot be placed on a drawing or forwarded to the workshop. To make sense on the drawing, the box is opened or spread-out onto a common plane which is the drawing sheet.

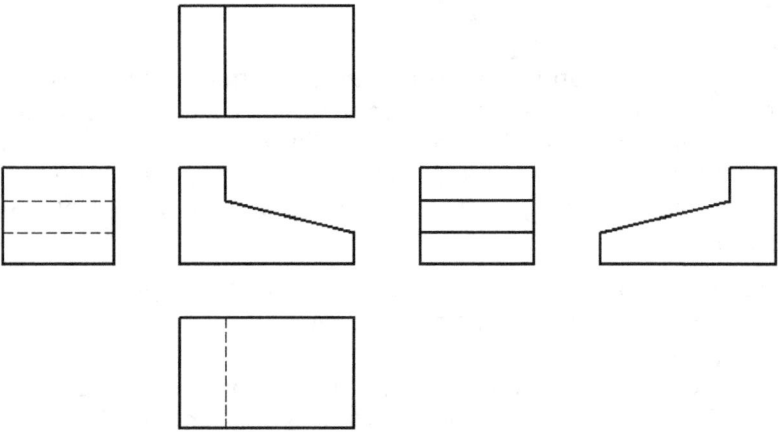

Figure 4.3

Figure 4.3 shows how the drawing would look like after cutting and spreading out the box.

Projection Systems:

Two methods of projection have been used to produce engineering drawings; Third Angle Projection is the preferred method stated in AS 1100 and is used throughout Australia in most drafting disciplines; however, some drafting disciplines still tend to use First Angle Projection. The difference between Third and First Angle Projection is the position of the Side, Top and Bottom Views in relation to the Front View. Until around 1890 all countries produced drawings in First Angle Projection, modern multi-national offices work entirely in Third Angle Projection.

Third Angle Projection:

The plane of projection lies between the observer and the object.

When the views are drawn, the Top View is located ABOVE the Front View, the Left Side View is located to the LEFT of the Front View, the Right View is located to the RIGHT of the Front View, and the Bottom View is located directly BELOW the Front View.

N.B. The views are drawn from where the object is being viewed. Viewed from the left and drawn on the left; viewed from on top and drawn on top.

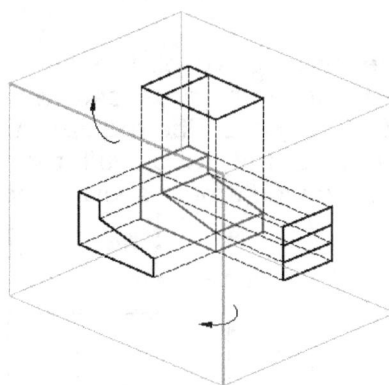

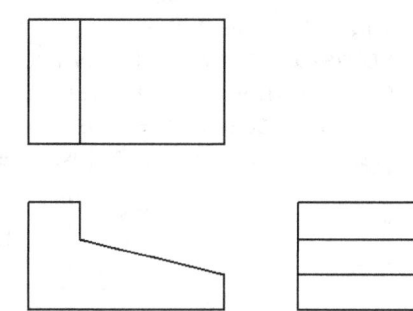

Figure 4. 4   *Figure 4. 5*

First Angle Projection:

The object lies between the observer and the plane.

When the views are drawn, the Top View is located BELOW the Front View, the Left Side View is located to the RIGHT of the Front View, the Right View is located to the LEFT of the Front View, and the Bottom View is located directly ABOVE the Front View. The location is exactly opposite to Third Angle Projection.

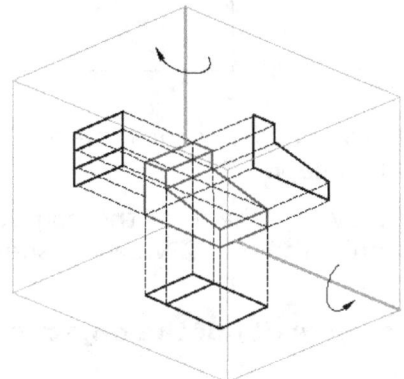

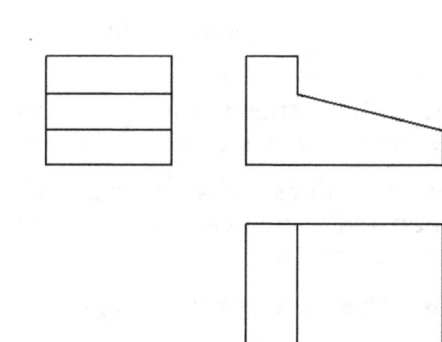

Figure 4. 6   *Figure 4. 7*

Projection Symbols:

To avoid misunderstanding, international projection symbols, as shown in Figure 4.8 and Figure 4.9, have been developed to distinguish between Third and First Angle projections on drawings.

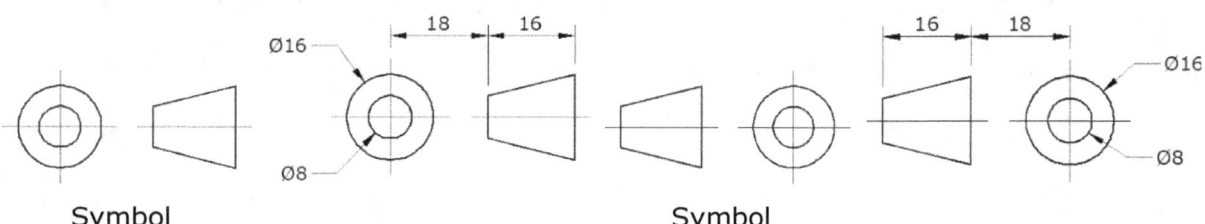

Symbol   Symbol

Figure 4.8 – Third Angle Projection   *Figure 4.9 – First Angle Projection*

Number of Views:

The number of views required depends on the complexity of the component; some drawings may require only one view with the width of the material shown under the Title while other components may require 5 or 6-views to fully describe the object. Figure 4.10 shows a complex cam that requires only 1-view to fully show all the features and dimensions; the thickness is constant, and Side View would only show a rectangle so the thickness can be placed below the Title.

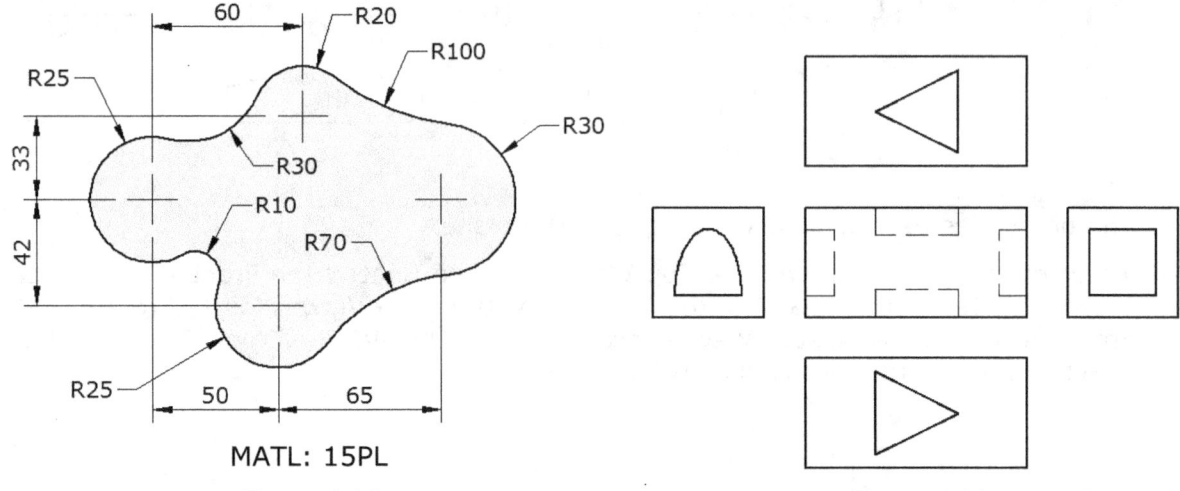

<table>
<tr><td>Figure 4.10</td><td>Figure 4.11</td></tr>
</table>

Figure 4.11 shows a simple hypothetical block with a series of different shape holes; however, 5-views are required to fully describe the shape.

Most detail drawings require 3-views to fully display the shape of the component and its dimensions however some components could only require 2-views, especially if it is symmetrical.

N.B. The number of views depends on the complexity of the object being drawn.

Skill Practice Exercises:

Skill Practice Exercise MEM09002-SP-0401
Identify the correct projection angle, or no projection used in the following examples.

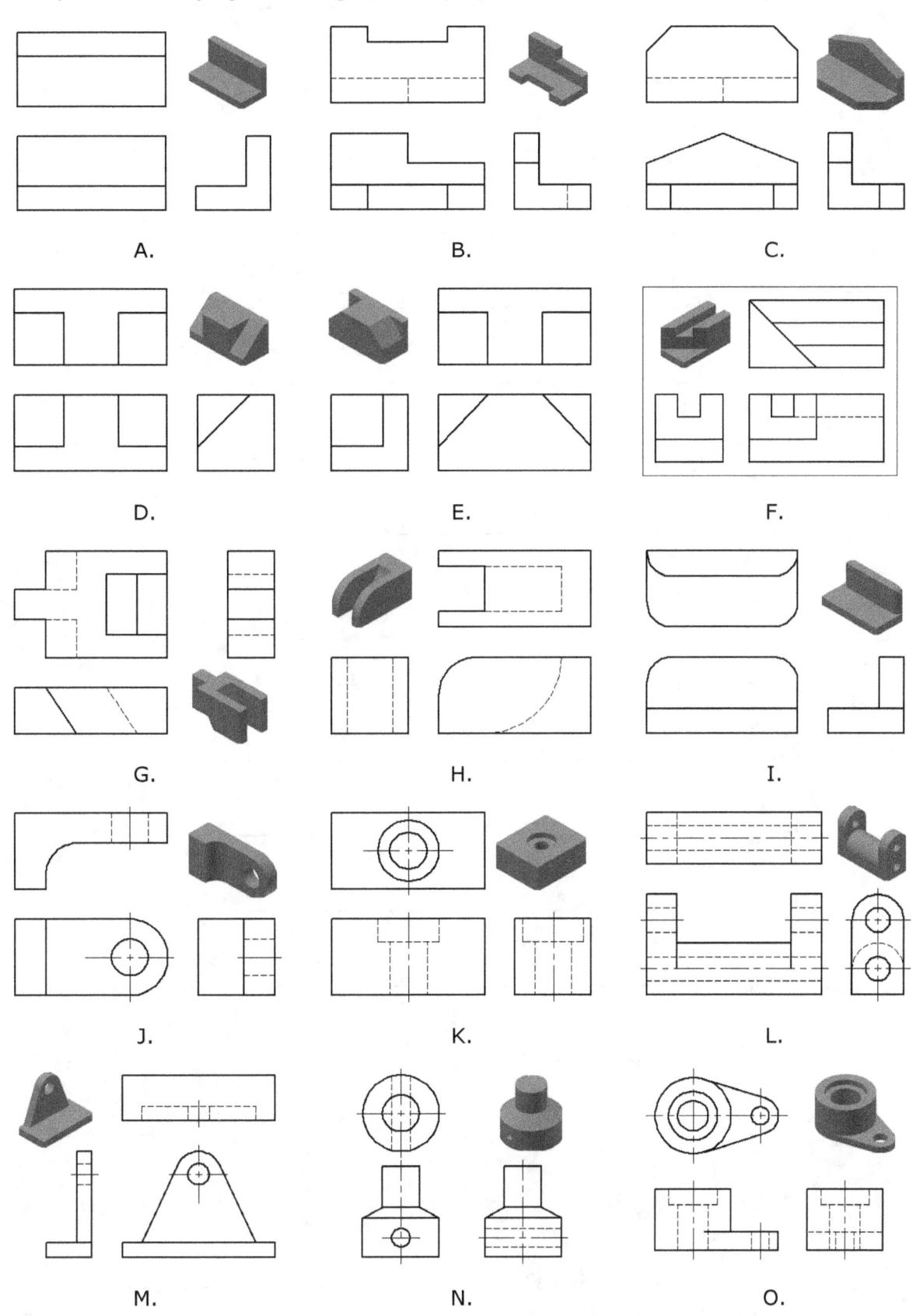

A.

B.

C.

D.

E.

F.

G.

H.

I.

J.

K.

L.

M.

N.

O.

MEM09002-SP-0401 Answer Sheet:

A. _____

B. _____

C. _____

D. _____

E. _____

F. _____

G. _____

H. _____

I. _____

J. _____

K. _____

L. _____

M. _____

N. _____

O. _____

Name: _____

Skill Practice Exercise MEM09002-SP-0402

Identify the correct projection angle used in the following examples. Answers are First Angle Projection, Third Angle Projection, Incorrect (First & Third Combined).

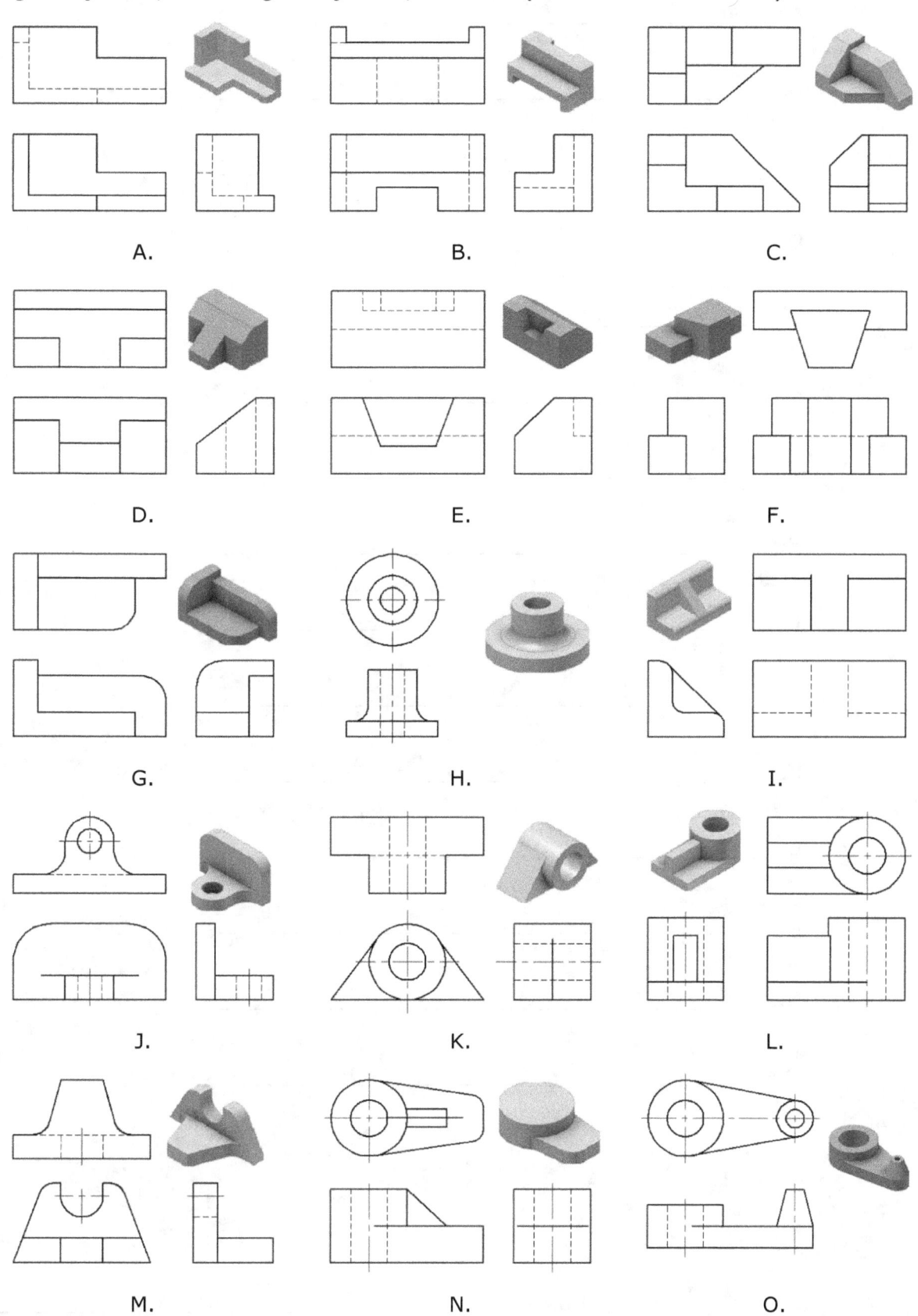

A.

B.

C.

D.

E.

F.

G.

H.

I.

J.

K.

L.

M.

N.

O.

MEM09002-SP-0402 Answer Sheet:

A. _____

B. _____

C. _____

D. _____

E. _____

F. _____

G. _____

H. _____

I. _____

J. _____

K. _____

L. _____

M. _____

N. _____

O. _____

Name: _____

Topic 5 – Units of Measurement:

Required Skills:
- Recognize SI units and their importance for measurement.
- Convert SI units to Imperial units.
- .Convert Imperial units to SI units

Required Knowledge:
- Understanding of the use of Imperial system of measurement.
- Basic calculation methods (addition, subtraction, multiplication, division).

History of Measurement:
The earliest recorded systems of weights and measures originate in the 3rd or 4th millennium BC. Even the very earliest civilizations needed measurement for purposes of agriculture, construction, and trade. Early standard units might only have applied to a single community or small region, with every area developing its own standards for lengths, areas, volumes and masses. Often such systems were closely tied to one field of use, so that volume measures used, for example, for dry grains were unrelated to those for liquids, with neither bearing any particular relationship to units of length used for measuring cloth or land.

The Cubit was unit of linear measure used by many ancient and medieval peoples. It may have originated in Egypt about 3000 BC; it thereafter became ubiquitous in the ancient world. The cubit, generally taken as equal to 457 mm (18 inches), was based on the length of the arm from the elbow to the tip of the middle finger and was considered the equivalent of 6 palms or 2 spans. In some ancient cultures it was about 531 mm (21 inches).

With development of manufacturing technologies, and the growing importance of trade between communities and ultimately across the Earth, standardized weights and measures became critical. Starting in the 18th century, modernized, simplified and uniform systems of weights and measures were developed, with the fundamental units defined by ever more precise methods in the science of metrology. The discovery and application of electricity was one factor motivating the development of standardized internationally applicable units.

Systems of Measurement:
There are two main systems for measuring distances and weight, the Imperial System of Measurement, and the Metric System of Measurement (SI Units). Most countries use the Metric System, which uses measuring units such as meters and grams and adds prefixes like kilo, milli and centi to count orders of magnitude. At present in the world, only 3-countries officially use the older Imperial system, the United States of America, Liberia and Myanmar where objects are measured in feet, inches, and pounds. The United Kingdom is officially metric however, imperial measures are still in use, especially for road distances, which are measured in miles.

SI Units (Metric System):
The first practical realisation of the metric system came in 1799, during the French Revolution, after the existing system of measures had become impractical for trade and was replaced by a decimal system based on the kilogram and the metre. The basic units were taken from the natural world. The unit of length, the metre, was based on the dimensions of the Earth, and the unit of mass, the kilogram, was based on the mass of a volume of water of one litre (a cubic decimetre). Reference copies for both units were manufactured in platinum and remained the standards of measure for the next 90 years.

The SI units of measurement have an interesting history. Over time they have been refined for clarity and simplicity. The 2-main daily used measurements being:

- The meter (m), or metre, was originally defined as 1/10,000,000 of the distance from the Earth's equator to the North Pole measured on the circumference through Paris. In modern terms, it is defined as the distance travelled by light in a vacuum over a time interval of 1/299,792,458 of a second.
- The kilogram (kg) was originally defined as the mass of a litre (i.e., of one thousandth of a cubic meter). It is currently defined as the mass of a platinum-iridium kilogram sample maintained by the Bureau International des Poids et Mesures in Sevres, France.

Length:

The most commonly used units of measurement are millimetres, centimetres, metres and kilometres. In engineering and architecture, the common unit of measure are the metre and millimetre while clothing and upholstery normally use centimetres.

Metre (m):

The metre is a standard measurement based on a metal bar on which a meter length has been marked to serve as the standard length of a meter which is realistically represented by the distance between two marks on an iron bar kept in Paris.

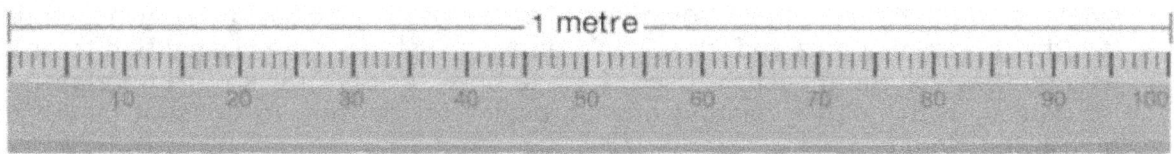

Convert metres to feet:
Multiply the distance (m) by 3.2808, e.g. to convert 3.25 m to feet
=3.25 x 3.2808 =**10.66′**.

Centimetre (cm):

A centimetre is a unit of length in the metric system, equal to one hundredth of a metre, centi being the SI prefix for a factor of. The centimetre was the base unit of length in the now deprecated centimetre–gram–second system of units. There are 100 centimetres in a metre.

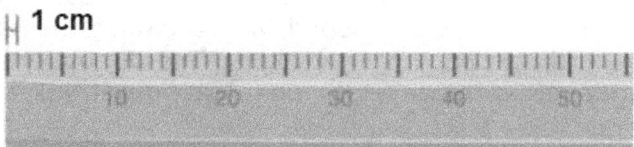

Millimetre (mm):

The millimetre or millimetre is a unit of length in the metric system, equal to one thousandth of a metre. There are one thousand millimetres in a metre and ten millimetres in a centimetre.

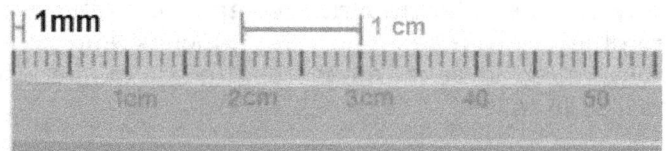

Convert millimetres to inches:
Multiply the distance (mm) by 0.0394, e.g. to convert 125 mm to inches
=125 x 0.0394 =**4.92″**.

Micrometre (μm):

The micrometre is commonly known as a micron and is an SI derived unit of length equalling 1×10^{-6} metre; that is, one millionth of a metre or 0.001 mm. The next smallest common SI unit is the nanometre, equivalent to one one-thousandth of a micrometre, or one billionth of a metre and used in the sub-atomic level of science & engineering.

Mass:

Weight is the measure of how heavy an object is. Weight is measured in standard customary units. The mass of an object is the amount of material it contains. Mass is measured in standard metric units.

For everyday purposes, when you're on the surface of the earth, the difference is not important. But if you measure something on another planet, its mass will be the same as it is on earth—but its weight will be different. (Weight depends on gravity, and gravity is different on other planets! This is why when you're floating in space, you're weightless. You still have mass, though,)

Tonne (t):

The tonne is the Standard International System of Units unit of mass and is equal to the mass of a cubic meter of pure water at the temperature of melting ice. 1 tonne equals 1000 kg.

Convert tonnes to ton:

> Multiply the metric mass (mm) by 0.9842, e.g. to convert 3.95 tonne to ton =3.95/1.1 =**3.89 ton**.

Kilogram (kg):

The kilogram is the Standard International System of Units unit of mass. It is defined as the mass of a particular international prototype made of platinum-iridium and kept at the International Bureau of Weights and Measures. It was originally defined as the mass of one litre (10^{-3} cubic meter) of pure water. 1,000 kilograms equals 1 tonne.

Convert kilogram to pounds:

> Multiply the metric mass (kg) by 2.205, e.g. to convert 56.48 kg to pounds =56.48 x 2.205 =**124.54 kg**.

Gram (g):

A gram is a unit of mass in the metric system defined as one thousandth (1×10^{-3}) of a kilogram. Originally, the gram was defined as a unit equal to the mass of one cubic centimetre of pure water.

Imperial System:

The British Imperial System evolved from the thousands of Roman, Celtic, Anglo-Saxon, and customary local units employed in the Middle Ages. Traditional names such as pound, foot, and gallon were widely used, but the values so designated varied with time, place, trade, product specifications, and dozens of other requirements. Early royal standards established to enforce uniformity took the name Winchester, after the ancient capital of Britain, where the 10th-century Saxon king Edgar the Peaceable kept a royal bushel measure and quite possibly others. Fourteenth-century statutes recorded a yard (perhaps based originally on a rod or stick) of 3 feet, each foot containing 12 inches, each inch equalling the length of three barleycorns (employed merely as a learning device since the actual standard was the space between two marks on a yard bar). Units of capacity and weight were also specified. In the late 15th century, King Henry VII reaffirmed the customary Winchester standards for capacity and length and distributed royal standards (physical embodiments of the approved units) throughout the realm. This process was repeated about a century later in the reign of Queen Elizabeth I. In the 16th century the rod (5.5 yards, or 16.5 feet) was defined (once again as a learning device and not as a standard) as the length of the left feet of 16 men lined up heel to toe as they emerged from church. By the 17th century usage and statute had established the acre, rod, and furlong at their present values (4,840 square yards, 16.5 feet, and 660 feet, respectively), together with other historic units. The several trade pounds in common use were reduced to just two: the troy pound, primarily for precious metals, and the pound avoirdupois, for other goods sold by weight.

Length:

The most commonly used units of measurement are miles, yards, feet and inches. Lesser units used in the Imperial system are furlongs, chains, rods, cables and links; in the ancient world the unit was a cubit which measured about 18-inches or 440 mm, Noah's Arc was reputed to measure 300 cubits long, 50 cubits wide and 30 cubits high.

Mile (m):

Historical Content: The Roman mile (mille passus, lit. "thousand paces"; abbr. m.p.; also milia passuum and mille) consisted of a thousand paces as measured by every other step as in the total distance of the left foot hitting the ground 1,000 times. The ancient Romans, marching their armies through uncharted territory, would often push a carved stick in the ground after each 1,000 paces. Well-fed and harshly driven Roman legionaries in good weather thus created longer miles. The distance was indirectly standardised by Agrippa's establishment of a standard Roman foot (Agrippa's own) in 29 BC, and the definition of a pace as 5 feet. An Imperial Roman mile thus denoted 5,000 Roman feet. Surveyors and specialised equipment such as the decempeda and dioptra then spread its use.

The modern mile is 1,760 yards or 5280 feet.

Convert miles to kilometres:
> Multiply the Imperial length (mile) by 1.609, e.g. to convert 257 miles to metres
> =257 x 1.609 = ***413.51 m***.

Yard (yd):

The yard is an English unit of length which measures 3 feet or 36 inches. The yard was the original standard adopted by the early English sovereigns and has been supposed to be founded upon the breadth of the chest of the Saxon race.

1 yard

Foot (ft or '):

The foot is the equivalent of 12 inches. The foot was a common unit of measurement throughout Europe. It often differed in length not only from country to country but from city to city. Because the length of a foot changed between person to person, measurements were not even consistent between two people, often requiring an average

1 foot

Convert feet to metres:
> Multiple the Imperial length (ft) by 0.3048, e.g. to convert 18 feet 6 inches to metres
> =12.5 x 3.048 = ***5.64 m***.

Inch (in or ")

The inch is 1/12th of a foot. History tells us that the inch was measured as the width of an average man's thumb at the base of the nail

1 inch

Convert inches to millimetres:
> Multiple the Imperial length (in) by 25.4, e.g. to convert 13.2 inches to millimetres
> =13.2 x 25.4 = ***335.28 mm***.

Mass:

Ton (t):

The British ton is the long ton, which is 2240 pounds, while the U.S. ton is the short ton which is 2000 pounds.

Convert tone to tonne:
> Multiple the Imperial mass (t) by 1.016, e.g. to convert 689 ton to tonne

=689 x 1.016 = **700.06 t**.

Hundredweight (cwt):
The English hundredweight is called the imperial hundredweight and is the equivalent to 8 stone, or 112 lbs.

Stone (st):
Stone, British unit of weight for dry products generally equivalent to 14 pounds.

Pound (lb):
The pound is a unit of mass used in the imperial, United States customary and other systems of measurement and is the equivalent of 16 ounces

Convert pound to kilogram:
Multiple the Imperial mass (lb) by 0.4536, e.g. to convert 56 pond to kilogram
=56 x 0.4536 = **25.40 kg**.

Ounce (oz):
The ounce is the smallest of weight or mass and is 1/16th of a pound.

Convert inches to millimetres:
Multiple the Imperial mass (oz) by 28.3495, e.g. to convert 19 oz to grams
=19 x 28.3495 = **538.64 g**

Temperature:

Temperature is a physical quantity that expresses hot and cold. It is the manifestation of thermal energy, present in all matter, which is the source of the occurrence of heat, a flow of energy, when a body is in contact with another that is colder.

Temperature is measured with a thermometer. Thermometers are calibrated in various temperature scales that historically have used various reference points and thermometric substances for definition. The most common scales are the Celsius scale (formerly called centigrade, denoted °C), the Fahrenheit scale (denoted °F), and the Kelvin scale (denoted K), the last of which is predominantly used for scientific purposes by conventions of the International System of Units (SI).

The lowest theoretical temperature is absolute zero, at which no more thermal energy can be extracted from a body. Experimentally, it can only be approached very closely, but not reached, which is recognized in the third law of thermodynamics.

The 4-temperature scales are Celsius, Fahrenheit, Kelvin and Rankine.

The formula to convert Celsius to Fahrenheit is:

(°C x 9/5) + 32

The formula to convert Fahrenheit to Celsius is:

(°F – 32) x 5/9

Celsius (° C):
The SI unit of measurement for temperature is Celsius range was originally defined by setting zero as the temperature at which water froze. Zero degrees C was later redefined as the temperature at which ice melts. The other point at which Celsius was set – 100 degrees Celsius – was defined as the boiling point of water.

Degrees Centigrade and degrees Celsius are the same thing. Degrees Celsius (invented by Anders Celsius) are sometimes called Centigrade, because the scale was defined between 0 and 100 degrees, hence centi-grade meaning a scale consisting of 1/100ths

Convert °C to °F:
Convert 36.5°C to Fahrenheit = (36.5 x 9/5) + 32 = 97.7°F

Fahrenheit (° F):
The Imperial unit of measurement for temperature is the Fahrenheit scale where water freezes at 32 degrees, and boils at 212 degrees. Boiling and freezing point are therefore

180 degrees apart. Normal body temperature is considered to be 98.6 °F (in real-life it fluctuates around this value). Absolute zero is defined as -459.67°F.

Convert °F to °C:

Convert 147°F to Celsius = (147 -32) x 5/9 = 63.89°C

Kelvin (° K):

Kelvin temperature scale, a temperature scale having an absolute zero below which temperatures do not exist. Absolute zero , or 0°K, is the temperature at which molecular energy is a minimum, and it corresponds to a temperature of −273.15° on the Celsius temperature scale . The Kelvin degree is the same size as the Celsius degree; hence the two reference temperatures for Celsius, the freezing point of water (0°C), and the boiling point of water (100°C), correspond to 273.15°K and 373.15°K, respectively. When writing temperatures in the Kelvin scale, it is the convention to omit the degree symbol and merely use the letter K. The temperature scale is named after the British mathematician and physicist William Thomson Kelvin , who proposed it in 1848. Another absolute temperature scale, the Rankine temperature scale , is used by some engineers.

Rankin (° R):

The Rankin scale is a thermodynamic (absolute) temperature scale. It is based around absolute zero. Rankine is similar to the Kelvin scale in that it starts at absolute zero and 0 °Ra is the same as 0 K but is different as a change of 1 °Ra is the same as a change of 1 °F (Fahrenheit) and not 1 °C (Celsius). Note that the abbreviation °R is ambiguous, as it can also refer to the Reaumur scale.

The Kelvin and Rankine temperature scales are defined so that absolute zero is 0 kelvins (K) or 0 degrees Rankine (°R). The Celsius and Fahrenheit scales are defined so that absolute zero is −273.15 °C or −459.67 °F.

Comparison Chart:

	SI Units		Imperial
Length	1 millimetre (mm)		0.03937 inch (in or ")
	1 centimetre (cm)		0.3937 inch (in or ")
	1 metre (m)		1.0936 yards (yd)
	1 kilometre (km)		0.6214 miles (m)
	25.4 mm		1 in
	0.348 m		12 in or 1 foot (ft)
	0.9144 m		3 ft or 1 yd
	1.6093 m		1 mile or 1760 yd
	1.853 km		1 nautical mile or 2025.4 yd
Area	1 sq cm (cm^2)	100 mm^2	0.1550 in^2
	1 sq m (m$^{2)}$	10,000 cm^2	1.1960 yd^2
	1 hectare (ha)	10,000 m^2	2.4711 acres
	1 sq km (km$^{2)}$	100 ha	.3861 m^2
	6.4516 m^2		1 sq in (in^2)
	0.0929 m^2		1 sq ft (ft^2)
	0.8361 m^2		1 sq yd (yd^2)
	4046.9 m^2		1 acre
	2.59 km^2		1 sq mile (mile^2)

Topic 5 - Units of Measurement

Mass			
	1 milligram (mg)		0.0154 grain
	1 gram (g)	1,000 mg	0.0353 ounce (oz)
	1 kilogram (kg)	1,000 g	2.2046 pound (lb)
	1 tonne (t)	1,000 kg	1.1023 short ton
	1 tonne (t)	1,000 kg	0.9842 long ton
	28.35 g		1 ounce (oz)
	0.45636 kg		1 pound (lb)
	6.3503 kg		1 stone
	50.802 kg		1 hundredweight (cwt)
	0.9072 t		1 short ton
	1.0160 t		1 long ton
Temperature	0° Celsius (C)		32.00° Fahrenheit (F)
	38° C		100.40° F
	-17.78° C		0° F
	37.77° C		100° F
	0° C	273.15° K	
	100° C	373.15° K	
		255.37° K	0° F
		0° K	-459.67° K

Skill Practice Exercises:

Skill Practice Exercise MEM09002-SP-0501

Refer to drawing HC145-58 and answer the following questions:

1. Convert 82° C to the Fahrenheit scale.

2. Convert 123° F to the Celsius scale.

3. Convert -56° C to Kelvin.

4. What instrument in used to measure hotness or coldness or a fluid?

5. Convert -12.5° C to the Fahrenheit scale.

6. Convert 24.3° F to the Celsius scale.

7. One day the minimum temperature in Marble Bar W.A. was recorded as 112.6° F. What was the temperature in degree Celsius at Marble Bar on that day?

8. Convert 148° C to the Fahrenheit scale.

9. Convert 1589 °F to the Celsius scale.

10. Convert 687° K to Rankin.

Name: _____

Topic 6 – Assembly Drawings:

Required Skills:

- Recognize components that fit together.
- Create a Material, Parts or Cutting List to match the Assembly Drawing.
- Determine overall dimensions and identify the different parts by cross-referencing.

Required Knowledge:

- The different types of Assembly Drawings.
- An understanding of Orthogonal Projection and the placement of associated views.
- A knowledge of mating parts.

General:

There are a number of drawing types associated with the mechanical engineering design process and include General Arrangement Drawings, Arrangement Drawings, Assembly Drawings, Detail Drawings and Fabrication Drawings.

General Arrangement Drawings

General Arrangement drawings show overall views of the equipment and provide all of the information to produce transportation, layout and installation drawings. The drawing includes a list of the arrangement drawings. The drawing includes overall dimensions, installation details, overall weight/mass, weights of sub systems, and service supply details.

The general arrangement drawing includes references to the design documents. The drawing often also identifies relevant internal and external contract numbers. The drawn separate assemblies and parts will be identified with leader lines to balloons or a numbering system which include the arrangement reference number linking to the list of arrangement drawings.

An example of a typical general arrangement drawing is shown in

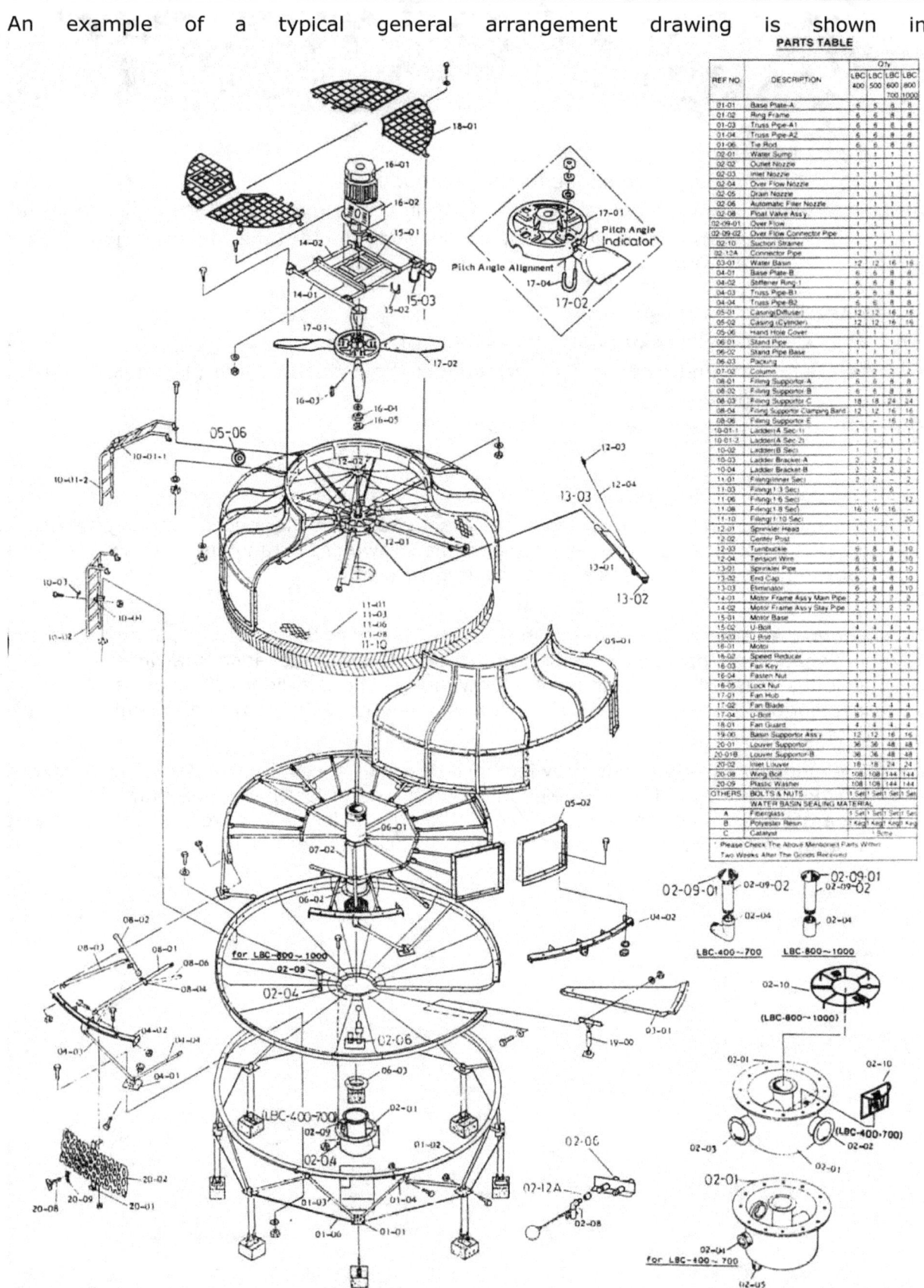

PARTS TABLE

REF NO.	DESCRIPTION	LBC 400	LBC 500	LBC 600 700	LBC 800 1000
01-01	Base Plate-A	6	6	8	8
01-02	Ring Frame	6	6	8	8
01-03	Truss Pipe-A1	6	6	8	8
01-04	Truss Pipe-A2	6	6	8	8
01-06	Tie Rod	6	6	8	8
02-01	Water Sump	1	1	1	1
02-02	Outlet Nozzle	1	1	1	1
02-03	Inlet Nozzle	1	1	1	1
02-04	Over Flow Nozzle	1	1	1	1
02-05	Drain Nozzle	1	1	1	1
02-06	Automatic Filter Nozzle	1	1	1	1
02-08	Float Valve Ass'y	1	1	1	1
02-09-01	Over Flow	1	1	1	1
02-09-02	Over Flow Connector Pipe	1	1	1	1
02-10	Suction Strainer	1	1	1	1
02-12A	Connector Pipe	1	1	1	1
03-01	Water Basin	12	12	16	16
04-01	Base Plate-B	6	6	8	8
04-02	Stiffener Ring-1	6	6	8	8
04-03	Truss Pipe-B1	6	6	8	8
04-04	Truss Pipe-B2	6	6	8	8
05-01	Casing(Diffuser)	12	12	16	16
05-02	Casing (Cylinder)	12	12	16	16
05-06	Hand Hole Cover	1	1	1	1
06-01	Stand Pipe	1	1	1	1
06-02	Stand Pipe Base	1	1	1	1
06-03	Packing	1	1	1	1
07-02	Column	2	2	2	2
08-01	Filling Supporter A	6	6	8	8
08-02	Filling Supporter B	6	6	8	8
08-03	Filling Supporter C	18	18	24	24
08-04	Filling Supporter Clamping Band	12	12	16	16
08-06	Filling Supporter E	–	–	16	16
10-01-1	Ladder(A Sec.1)	1	1	1	1
10-01-2	Ladder(A Sec.2)	1	1	1	1
10-02	Ladder(B Sec)	1	1	1	1
10-03	Ladder Bracket-A	2	2	2	2
10-04	Ladder Bracket-B	2	2	2	2
11-01	Filling(3inner Sec)	2	2	–	2
11-03	Filling(1 3 Sec)	–	–	6	–
11-06	Filling(1 6 Sec)	–	–	–	12
11-08	Filling(1 8 Sec)	16	16	16	–
11-10	Filling(1 10 Sec)	–	–	–	20
12-01	Sprinkler Head	1	1	1	1
12-02	Center Post	1	1	1	1
12-03	Turnbuckle	6	8	8	10
12-04	Tension Wire	6	8	8	10
13-01	Sprinkler Pipe	6	8	8	10
13-02	End Cap	6	8	8	10
13-03	Eliminator	6	8	8	10
14-01	Motor Frame Ass'y Main Pipe	2	2	2	2
14-02	Motor Frame Ass'y Stay Pipe	2	2	2	2
15-01	Motor Base	1	1	1	1
15-02	U-Bolt	4	4	4	4
15-03	U-Bolt	4	4	4	4
16-01	Motor	1	1	1	1
16-02	Speed Reducer	1	1	1	1
16-03	Fan Key	1	1	1	1
16-04	Fasten Nut	1	1	1	1
16-05	Lock Nut	1	1	1	1
17-01	Fan Hub	1	1	1	1
17-02	Fan Blade	4	4	4	4
17-04	U-Bolt	8	8	8	8
18-01	Fan Guard	4	4	4	4
19-00	Basin Supporter Ass'y	12	12	16	16
20-01	Louver Supporter-A	36	36	48	48
20-01B	Louver Supporter-B	36	36	48	48
20-02	Inlet Louver	18	18	24	24
20-08	Wing Bolt	108	108	144	144
20-09	Plastic Washer	108	108	144	144
OTHERS	BOLTS & NUTS	1 Set	1 Set	1 Set	1 Set
	WATER BASIN SEALING MATERIAL				
A	Fiberglass	1 Set	1 Set	1 Set	1 Set
B	Polyester Resin	1 Kg	1 Kg	1 Kg	1 Kg
C	Catalyst			Bottle	

Please Check The Above Mentioned Parts Within
Two Weeks After The Goods Received

Figure 6. 1 and shows an exploded view of a Cooling Tower with the components cross referenced to part and reference drawing numbers attached. Another example could be a roller conveyor system comprising a number of conveyors with independent drives and guards. In the maritime industry, the layout of ship's decks and side view are called General Arrangement drawings. No referencing to part or drawing numbers are given, only the names of the compartment, spaces and equipment as shown in Figure 6. 2 – Maritime General Arrangement Drawing.

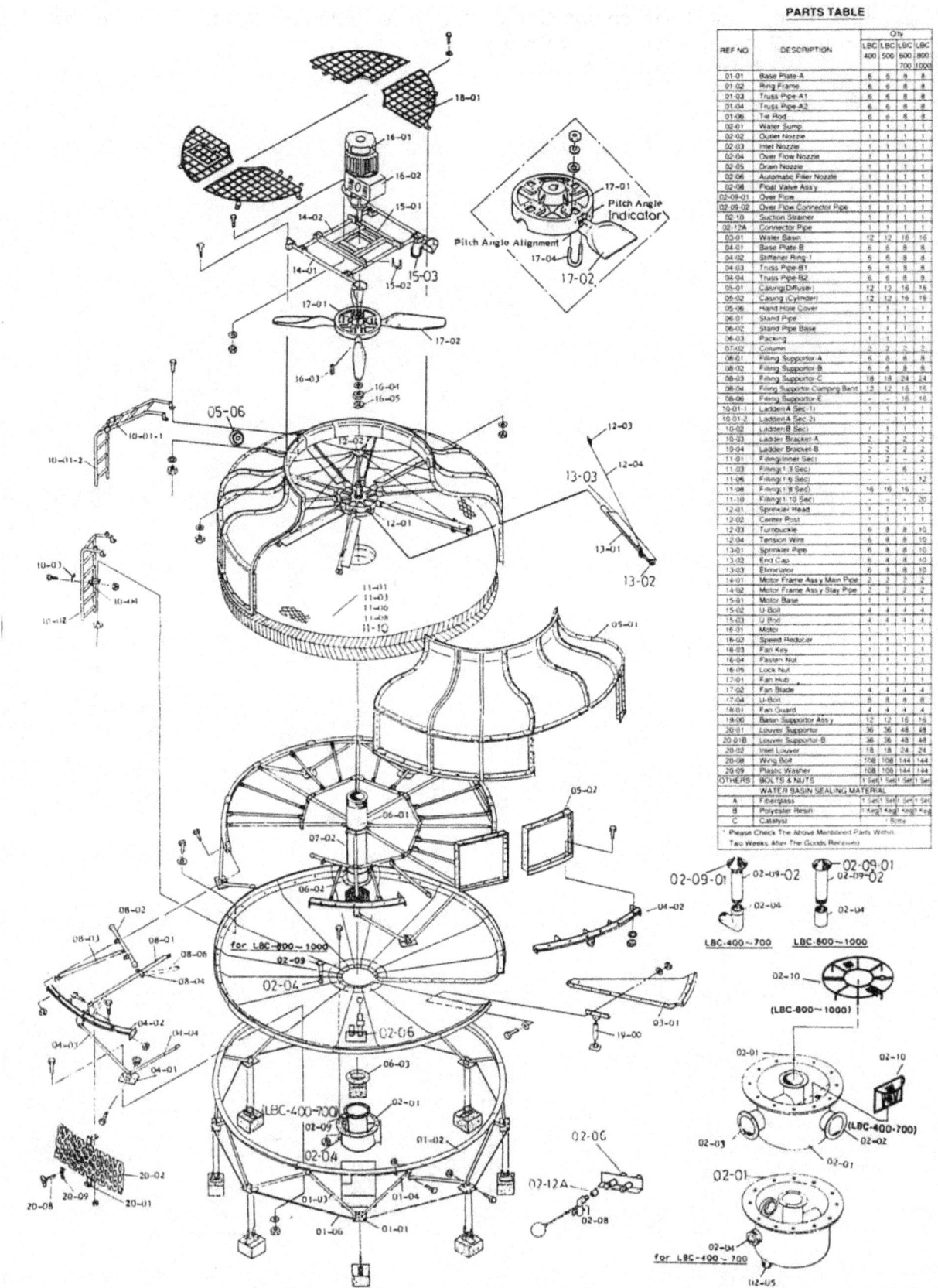

PARTS TABLE

REF NO	DESCRIPTION	Qty LBC 400	LBC 500	LBC 600 700	LBC 800 1000
01-01	Base Plate-A	6	6	8	8
01-02	Ring Frame	6	6	8	8
01-03	Truss Pipe-A1	6	6	8	8
01-04	Truss Pipe-A2	6	6	8	8
01-06	Tie Rod	6	6	8	8
02-01	Water Sump	1	1	1	1
02-02	Outlet Nozzle	1	1	1	1
02-03	Inlet Nozzle	1	1	1	1
02-04	Over Flow Nozzle	1	1	1	1
02-05	Drain Nozzle	1	1	1	1
02-06	Automatic Filler Nozzle	1	1	1	1
02-08	Float Valve Assy	1	1	1	1
02-09-01	Over Flow	1	1	1	1
02-09-02	Over Flow Connector Pipe	1	1	1	1
02-10	Suction Strainer	1	1	1	1
02-12A	Connector Pipe	1	1	1	1
03-01	Water Basin	12	12	16	16
04-01	Base Plate-B	6	6	8	8
04-02	Stiffener Ring-1	6	6	8	8
04-03	Truss Pipe-B1	6	6	8	8
04-04	Truss Pipe-B2	6	6	8	8
05-01	Casing (Diffuser)	12	12	16	16
05-02	Casing (Cylinder)	12	12	16	16
05-06	Hand Hole Cover	1	1	1	1
06-01	Stand Pipe	1	1	1	1
06-02	Stand Pipe Base	1	1	1	1
06-03	Packing	1	1	1	1
07-02	Column	2	2	2	2
08-01	Filling Supporter-A	6	6	8	8
08-02	Filling Supporter-B	6	6	8	8
08-03	Filling Supporter-C	18	18	24	24
08-04	Filling Supporter Clamping Band	12	12	16	16
08-06	Filling Supporter-E	-	-	16	16
10-01-1	Ladder-A Sec-1	1	1	1	1
10-01-2	Ladder-A Sec-2	-	1	1	1
10-02	Ladder-B Sec	1	1	1	1
10-03	Ladder Bracket-A	2	2	2	2
10-04	Ladder Bracket-B	2	2	2	2
11-01	Filling (liner Sec)	2	2	-	2
11-03	Filling (1-3 Sec)	-	-	6	-
11-06	Filling (1-6 Sec)	-	-	-	12
11-08	Filling (1-8 Sec)	16	16	16	-
11-10	Filling (1-10 Sec)	-	-	-	20
12-01	Sprinkler Head	1	1	1	1
12-02	Center Post	1	1	1	1
12-03	Turnbuckle	8	8	8	10
12-04	Tension Wire	8	8	8	10
13-01	Sprinkler Pipe	8	8	8	10
13-02	End Cap	8	8	8	10
13-03	Eliminator	8	8	8	10
14-01	Motor Frame Assy Main Pipe	2	2	2	2
14-02	Motor Frame Assy Stay Pipe	2	2	2	2
15-01	Motor Base	1	1	1	1
15-02	U-Bolt	4	4	4	4
15-03	U Bolt	4	4	4	4
16-01	Motor	1	1	1	1
16-02	Speed Reducer	1	1	1	1
16-03	Fan Key	1	1	1	1
16-04	Fasten Nut	1	1	1	1
16-05	Lock Nut	1	1	1	1
17-01	Fan Hub	1	1	1	1
17-02	Fan Blade	4	4	4	4
17-04	U-Bolt	8	8	8	8
18-01	Fan Guard	4	4	4	4
19-00	Basin Supporter Assy	12	12	16	16
20-01	Louver Supporter	36	36	48	48
20-01B	Louver Supporter-B	36	36	48	48
20-02	Inlet Louver	18	18	24	24
20-08	Wing Bolt	108	108	144	144
20-09	Plastic Washer	108	108	144	144
OTHERS	BOLTS & NUTS	1 Set	1 Set	1 Set	1 Set
	WATER BASIN SEALING MATERIAL				
A	Fiberglass	1 Set	1 Set	1 Set	1 Set
B	Polyester Resin	1 Keg	1 Keg	1 Keg	1 Keg
C	Catalyst		1 Bottle		

* Please Check The Above Mentioned Parts Within Two Weeks After The Goods Received.

Figure 6. 1 - General Arrangement Drawing

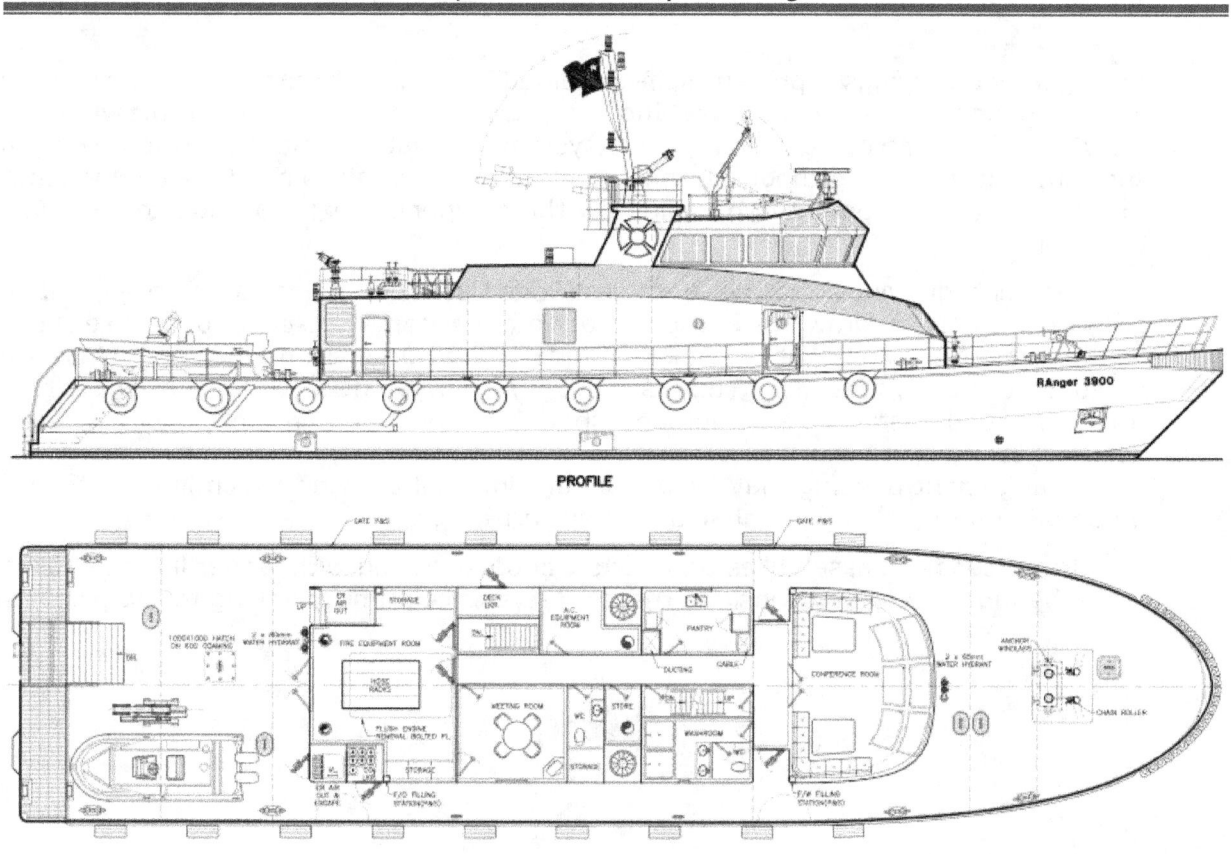

PROFILE

MAIN DECK PLAN

Figure 6. 2 – Maritime General Arrangement Drawing

Arrangement Drawing

Arrangement drawings represent self-contained units used to make up the system drawn on the general arrangement drawing. Examples of arrangement drawings include drawings of assembled conveyers, drive systems, elevating units etc. The drawing should show in, at least three orthographic views, clear details to show all of the components used to make up the equipment items and how the component parts are located and fastened together.

Arrangement drawings include a table (Part List) identifying assemblies, fabrication drawings, detail drawings and proprietary items used to make up the equipment. Arrangement drawings include overall dimension, the weight/mass of the equipment drawn, the lifting points. All information needed to construct, test, lift, transport, and install the equipment should be provided in notes or as referenced documents.

The arrangement drawing may be a standard internal drawing which is repeatedly called up on different system general arrangement drawings.

The drawn separate assemblies and parts will be identified with leader lines to balloons or a numbering system which include the item reference number linking to the parts list.

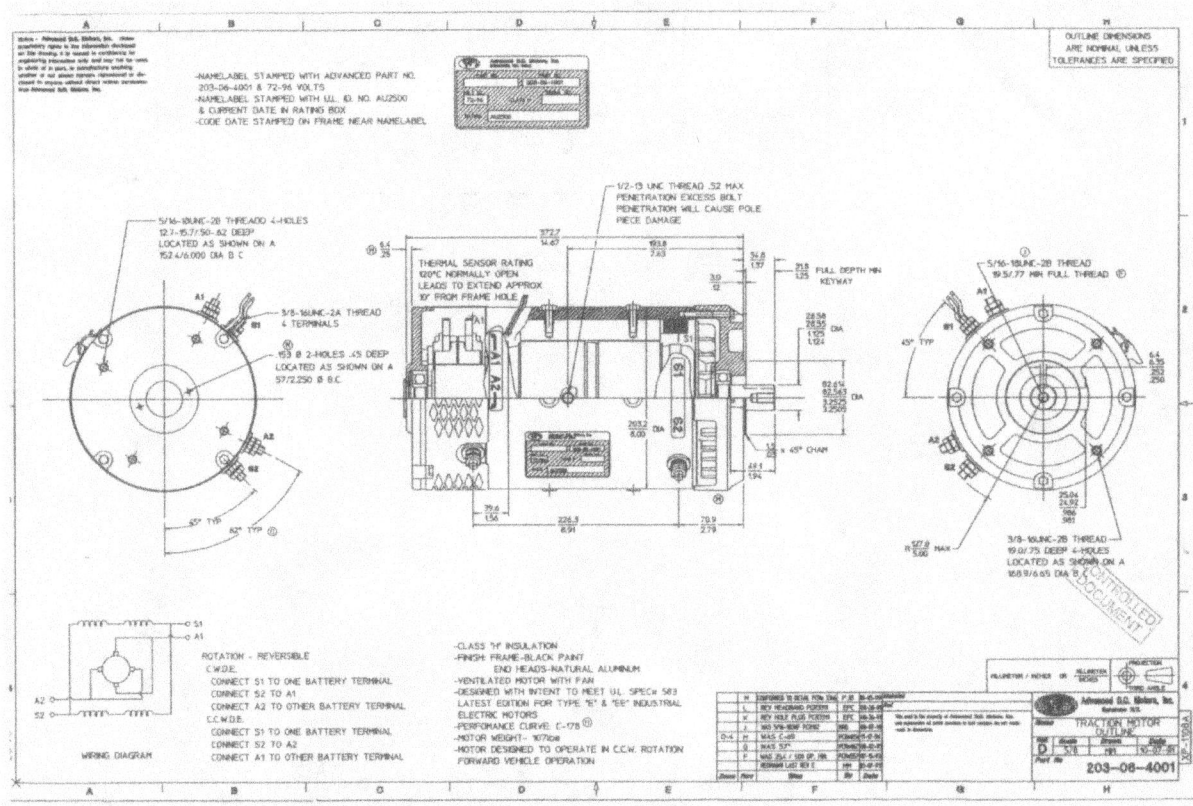

Figure 6. 3 – Arrangement Drawing

Assembly Drawings

The assembly /sub-assembly drawings are drawings of discrete sub-systems showing in some detail how the component items fit together. Typical assembly drawings include gearbox drawings, roller drawings, guard system drawings.

The assembly drawing will generally include at least three orthographic views with sections as needed to clearly show all of the details and their relative positions. Overall and detail dimensions will be shown. The weight/mass of the assembly/sub-assembly will be noted. The drawing will include a parts list identifying all of the component details with quantities and materials and supply details. The assembly drawing will include a list of reference drawings and notes identifying the relevant codes and specifications and testing requirements.

The drawn separate items will be identified with leader lines to balloons or a numbering system which include the item reference number linking to the parts list.

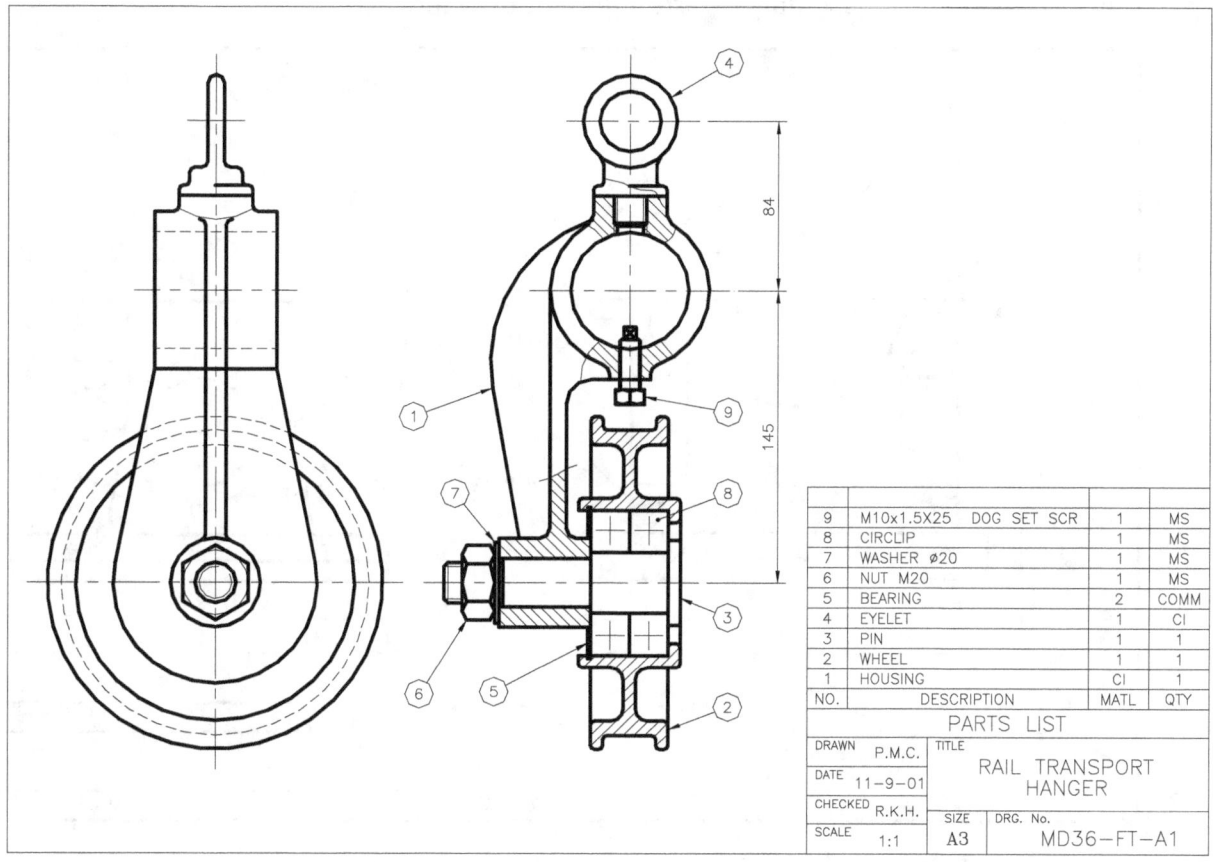

9	M10x1.5X25 DOG SET SCR	1	MS
8	CIRCLIP	1	MS
7	WASHER ⌀20	1	MS
6	NUT M20	1	MS
5	BEARING	2	COMM
4	EYELET	1	CI
3	PIN	1	1
2	WHEEL	1	1
1	HOUSING	CI	1
NO.	DESCRIPTION	MATL	QTY

PARTS LIST

DRAWN	P.M.C.	TITLE	RAIL TRANSPORT HANGER		
DATE	11-9-01				
CHECKED	R.K.H.				
SCALE	1:1	SIZE A3	DRG. No. MD36-FT-A1		

Figure 6. 4 - Assembly Drawing

Detail Drawings

All individual items required to produce mechanical equipment need to be described in some detail to ensure that they are manufactured in accordance with the designers requirements. Proprietary items are selected from technical data sheets obtained from manufacturer /supplier. Items manufactured specifically for the application need to be made to detail drawings which include the geometry, material, heat treatment requirements, surface texture, size tolerances, geometric tolerances etc.

The detail drawing should include all of the necessary information to enable procurement, manufacture and should identify all of the relevant codes and standards. The item weight/mass should also be included for reference.

Depending on the level of detail, a detail drawing can comprise one drawing on a sheet or a number of separate drawings on one sheet. It is sometimes possible to combine the detail drawings onto the assembly drawing. The detail drawing must cross reference, both ways, to the parent assembly or arrangement drawing.

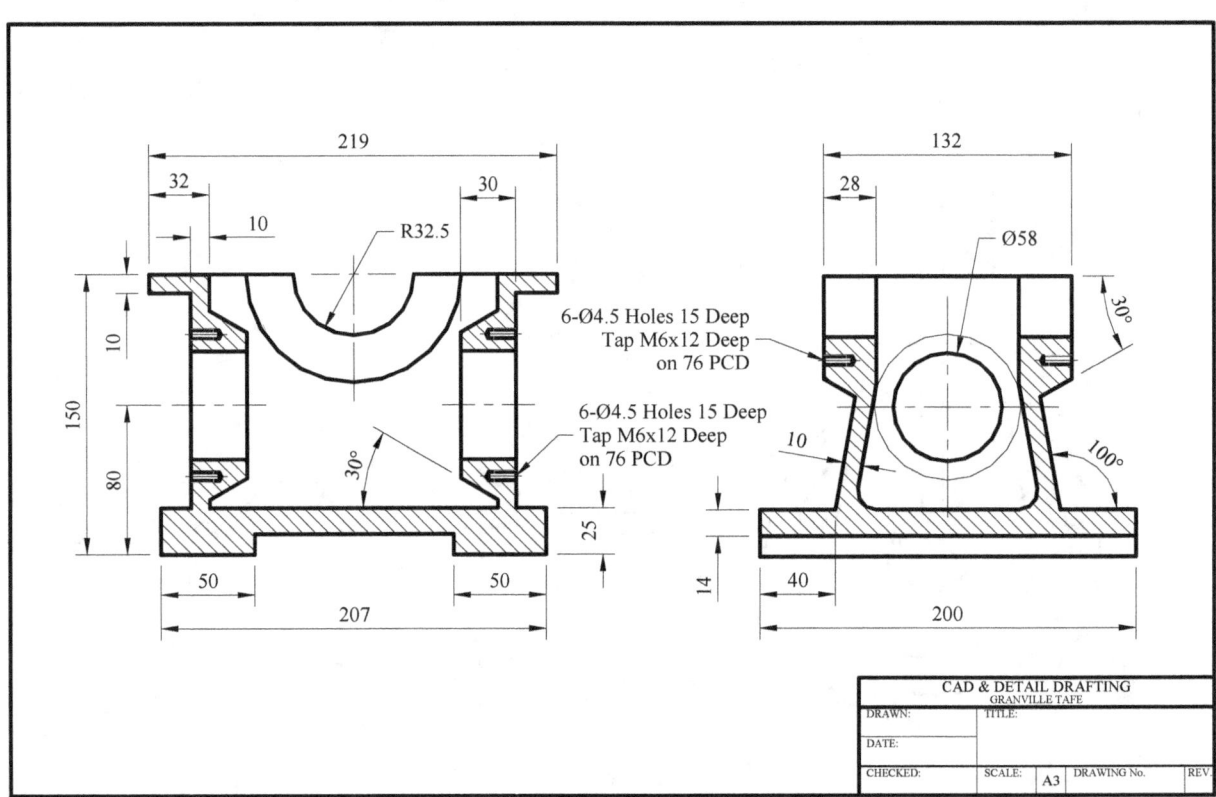

Figure 6. 5 - Detail Drawing

Fabrication Drawings

The fabrication drawing is a specific type of detail drawing. Some fabrication drawings are virtually assembly drawing e.g. when a number of items are assembled together as a fabrication. The fabrication drawing generally includes a material parts list identifying all of the materials used to build up the fabrication. All weld details are included using the standard symbolic representation of welds as shown in BS EN 22553. All of the materials should be identified in accordance with the relevant standards and codes.

The fabrication drawing should clearly describe in notes or in referenced documents the heat treatment and stress relieving requirements prior to, during and following the completion of the fabrication processes. The dimensions and relevant linear and geometric tolerances should be indicated.

A fabrication drawing sometimes only includes the fabrication details, the final machining details are then shown on a separate drawing. It is equally acceptable to show all manufacturing information on one drawing.

The items used to make up the fabrication will be identified with leader lines to balloons which include the item reference number linking to the parts list. The listed items on a fabrication drawing do not identify items which can be disassembled, as on assembly and arrangement drawings. The numbering system should reflect this difference. Methods of numbering items on fabrication drawings include using lower case alphabet letters e.g a,b,c or optionally as sub units of the fabrication item number e.g 1/1, 1/2 1/3 ... or 1/a , 1/b, 1/c...

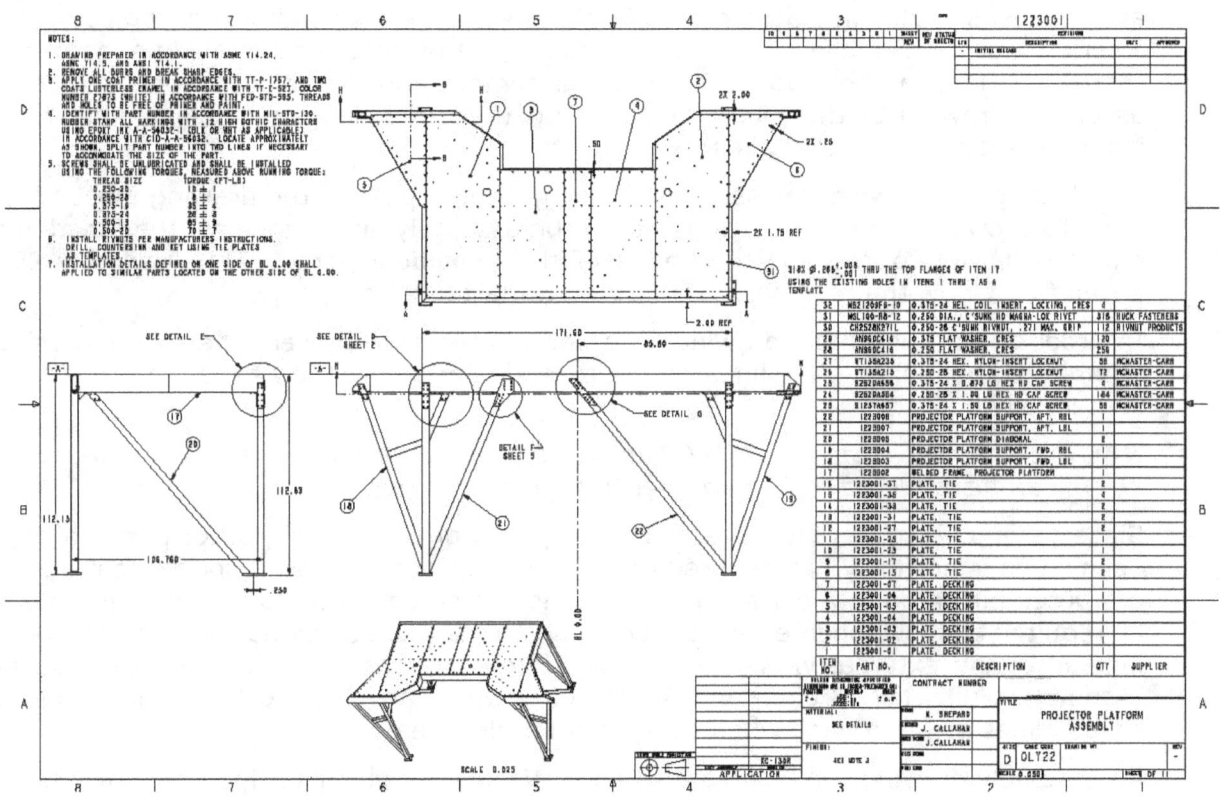

Figure 6. 6 - Fabrication Drawing

Item Identification

The method of identifying the parts must be clear and unambiguous. The equipment as represented on the general arrangement drawing and the sub-assemblies as shown on the arrangement and assembly drawing should be clearly identified with plant item numbers. The relevant drawing numbers are obtained by reference to the plant items list. Plant items are annotated by leader lines to a double balloon.

Typically a conveyor may have a plant item number e.g.H1040 and be shown on a drawing e.g. drawing number A0 12500.

The detail drawings are sub items of the arrangement drawings and are identified on the arrangement and assembly drawings. Typically an item say a conveyor frame may be identified from the conveyor plant item number e.g. H1040/3. Optionally it may be identified using the arrangement drawing number e.g. A0 12500 /3. The frame will also have a discrete detail drawing number e.g A2 12503.

The fabricated items which are based on sub-parts welded together should be identified as details but the individual sub-parts should be identified in a different way to avoid ambiguity. One option is to number the fabricated sub-parts alphabetically e.g. a, b, c ...or as a combination of the fabrication detail number and the part number i.e. 3/a, 3/b.... These sub-parts do not need to be identified as separate parts because following fabrication they will not exist as separate parts. If the sub-parts are complicated shapes or machined items and they cannot be described in sufficient detail on the fabrication drawing they should be drawn as separate detail drawings but still identified as sub-parts of the fabrication detail.

Assembly Drawings:

As can be seen in Figure 6. 4 - Assembly Drawing, an Assembly Drawing shows the relative positions of the different parts. The Assembly drawing also proves the different parts fit together without designed interference and with the correct clearances for moving parts.

Assembly Drawings include preliminary design drawings and layouts, piping plans, unit assemblies, installation diagrams and final drawings. An Assembly Drawing can consist of a series of sub-assemblies; e.g. the gear box is part of the drive system in a motor vehicle, therefore the gear box would be a sub-assembly and would also appear in the final assembly drawing of the motor vehicle with the other sub-assemblies (engine, clutch, steering, suspension and brakes).

In selecting the views for a assembly drawing, the purpose of the drawing must be kept in mind to show how the parts fit together in the assembly and to suggest the function of the entire unit, not to describe the shape of the individual parts. The Assembly Drawing purports to show the *relationships* of different parts, not *shapes*.

In producing an assembly drawing, one line is used to represent the mating surfaces of different components. The lines can be extended or shortened as required; the assembly is gradually built up.

NB: *The different parts are **NEVER** drawn as individual parts then moved into place as errors can be made ad the 2 parts not fit together correctly*.

Since assemblies generally have parts fitting into or overlapping other parts, hidden line delineation is normally not required unless it is required to show a special feature such as a tapered pin through 2 parts. If the assembly is so complicated that hidden lines would be required to show the internal detail clearly, one or more sectional views should be drawn instead of the external views. Any type of sectional view can be used to describe the assembly; Full, Half, Broken and Removed sections are the most common types used. Hidden lines are only used when necessary for clearness.

An assembly drawing can be created using CAD software by inserting pre-drawn parts into the drawing and then positioning them as required, or, the parts drawn by extending and/or trimming lines to create the assembly.

Features of an Assembly Drawing:

An Assembly Drawing consists of the views (normally 2 or 3), Material/Parts/Cutting List, cross referencing (called balloons) to the Material (or Parts or Cutting List) and notes covering the manufacturing processes required for the assembly; the drawing may also show the overall dimensions to indicate the space required for the assembly if it is to be shipped or fit into a specific area. Dimensions between centre distances may also be included in the case of belt or chain drive systems to assist the technician in building the equipment.

Parts List:

The Parts List is also known as a Material List or Cutting List depending on the drafting discipline. Parts Lists are commonly used in mechanical disciplines while Material and Cutting Lists are generally used in the construction disciplines. The list is usually placed on the drawing immediately above the Title Block but can also be placed on a separate sheet, especially if there are many components or if the company has a special estimating section that does not require the drawing, only a list of the components.

The Parts List can consist of the basic information (identification number, part name or description, material and quantity) or contain other specific and important information (specification numbers, catalogue numbers, drawing numbers, remarks, stock number, manufacturer and/or supplier) displayed in columns.

The typical layout of a Parts List used in TAFE is shown below:

NO.	DESCRIPTION	MATL	QTY
3	PIN	CRS	15
2	BOLT HEX HD M12x1x50	ALAL	8
1	75x10FLx1500	MS	1
NO.	DESCRIPTION	MATL	QTY
	PARTS LIST		

Balloons or Cross Referencing:

A balloon is a circle that contains a single number, which is connected with a leader line pointing to the part within the assembly. Balloon Guidelines include:

- All balloons on a drawing must be the same size.
- Balloons should be grouped together in an easy to read pattern.
- Balloon numbers must correspond to the item numbers in the Parts List.
- Balloons should not have horizontal or vertical leader lines.

The diameter of the balloon is a direct proportion to the height of the text (2.4xH). The leader should have a short horizontal reference line before the leader, or, the leader end at the balloon and pointing to the centre. The text should be placed (or justified) about the centre of the circle.

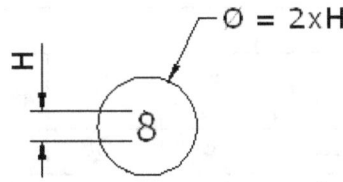

Dimensions:

As a rule, dimensions are not given on assembly drawings since they are given on the detail individual drawings. If dimensions are provided, they are limited to some function of the object as a whole, such as the overall height, width and length of the assembly, the maximum/minimum opening between two components within the assembly, or, the centre-to-centre distances between gears, pulleys and sprockets. Dimensions would be required when the assembly requires surfaces to be machined after the components have been assembled.

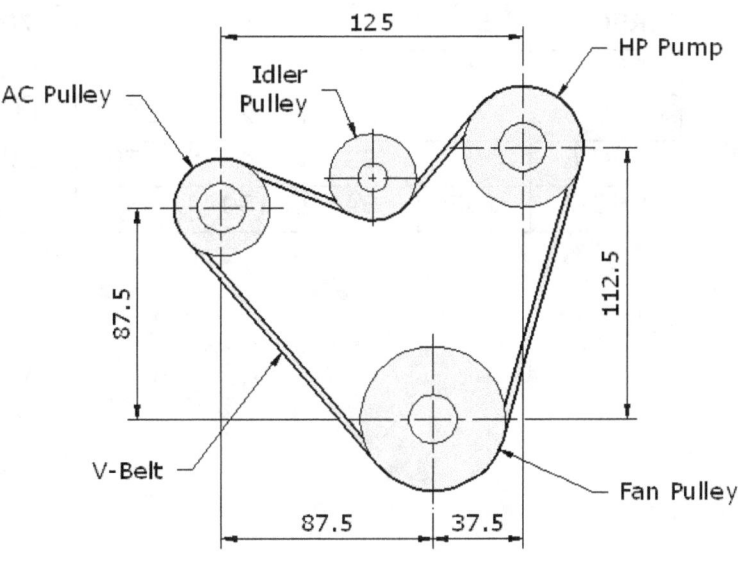

Figure 6. 7

Tabulation:

Tables are used in outline assembly drawings to give the general idea of the exterior shape of a machine or structure and consist of only the principle dimensions as shown in Figure 6. 8 referring to the table showing those dimensions. When the drawing is made for catalogues or other illustrative purposes, the dimensions are often omitted.

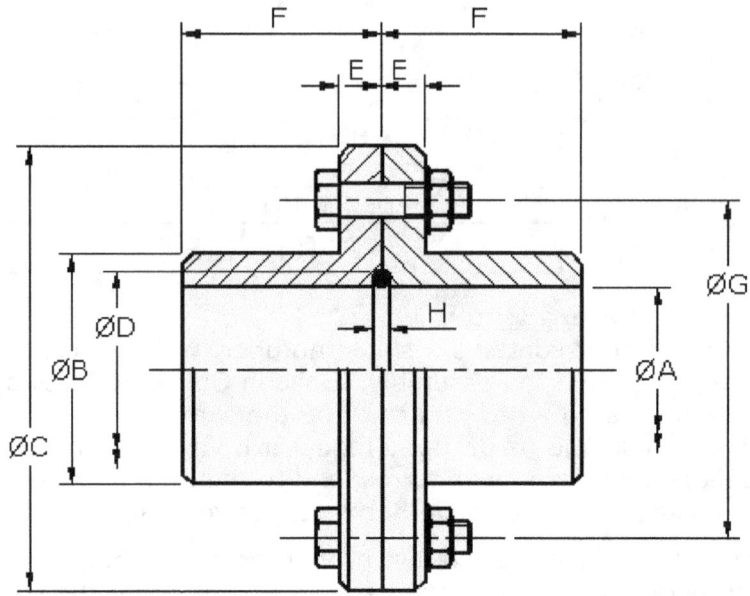

Figure 6. 8

Type	Dimensions								
	A max	A min	B	C	D	E	F	G	H
RSC-1	42	15	70	146	58	12	56	108	6
RSC-2	48	21	82	171	76	17	61	127	6
RSC-3	58	21	97	198	76	17	68	145	6
RSC-4	70	21	117	216	110	17	76	167	8
RSC-5	78	25	127	254	110	30	68	190	8
RSC-6	85	28	147	279	135	30	100	213	8
RSC-7	105	34	180	330	160	30	117	255	8

How Do The Parts Fit Together?

The draftsperson should be competent in being able to identify from the engineer's sketches, the way the different components fit together to form the assembly. The secret is to look for similar or matching dimensions and features including:

Threads: A shaft could have a threaded end which screws onto a mating pulley or fastening.   e.g. **M10x0.75**

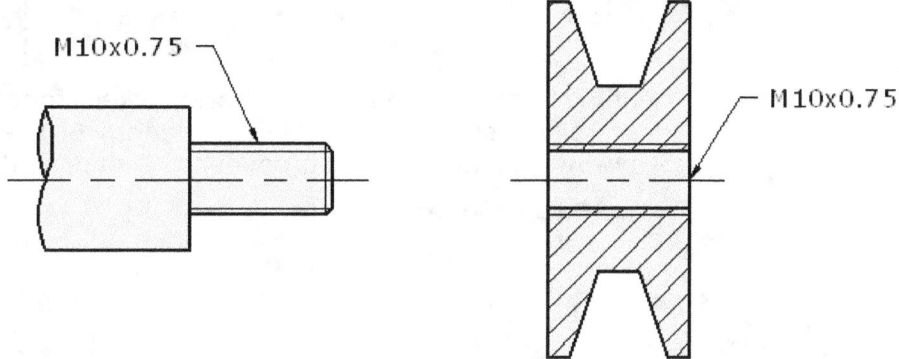

Shape: A component could have a dovetailed feature which would over another matching dovetail.

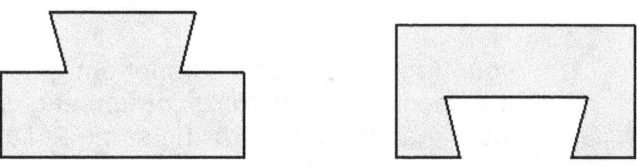

Taper   or   A taped face on one object will almost certainly fit against the face of
Angle: another angled surface. eg. **16°.**

In the above example the two angular dimensions form the same angle to the horizontal, the 16° is given from the horizontal while the 74° is given from the vertical; both add up to 90°.

Dimensions: A dimension on one part will probably match up to the same dimension on another part even though the parts may have different shapes.

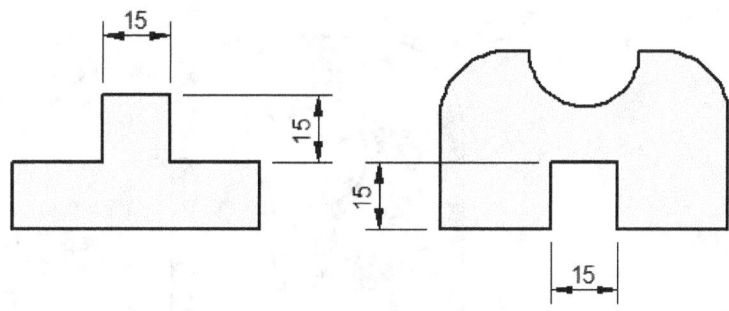

Toleranced Dimensions:	Although the toleranced dimensions of two mating parts may not have the same upper and lower values, they will be very similar with the same basic size.

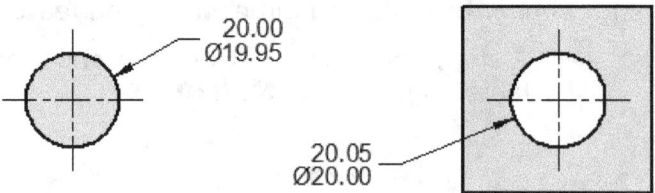

Keys and Keyways:	Many shafts and pulleys etc are locked together by keys which are generally a square or rectangular length of metal. On the drawing the keyway is identified by a groove in the shaft and pulley.

eg. **10x3 Keyway**.

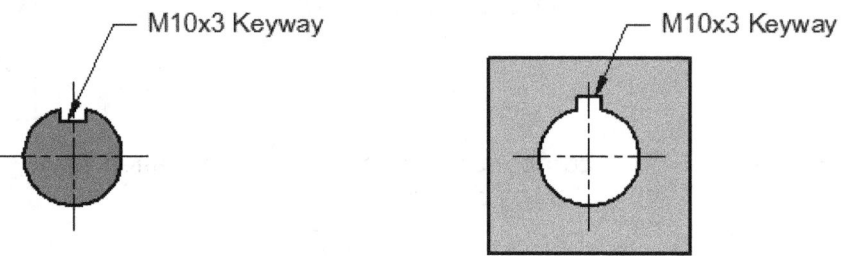

Holes:	Mounting holes in one component generally mate with another set of holes in a second and/or third component. eg. **6-Ø16 Holes on 150PCD** that could match holes with the same PCD.

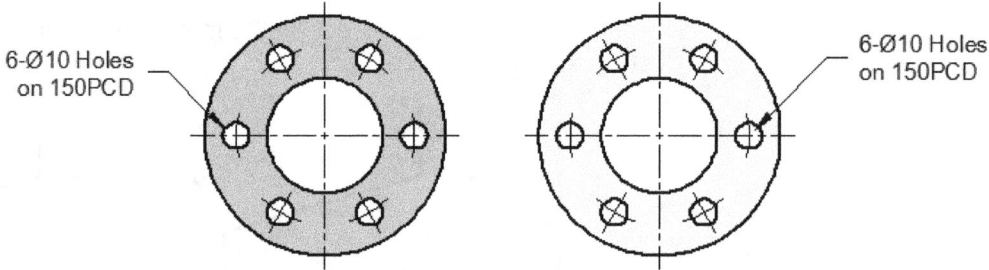

Fastenings:	A fastening will probably be threaded which will screw into a mating hole, eg. **M10** fastening and **Ø10** Holes with the identical **125 PCD**. The shape of the head of the fastening may also determine its location with a part, eg. an M10x30 Socket Head Cap Screw would fit into a counterbored hole on one part which would be identified **Drill Ø10 C'bore Ø16x10 deep** with a matching PCD.

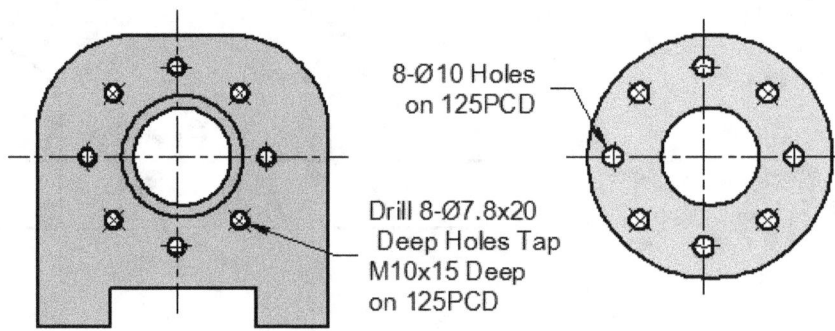

Skill Practice Exercises:

Use the following Assembly and Detail Drawings to determine the answers for Skill Practice Exercises MEM09002-SP-0601 and MEM09002-SP-0602.

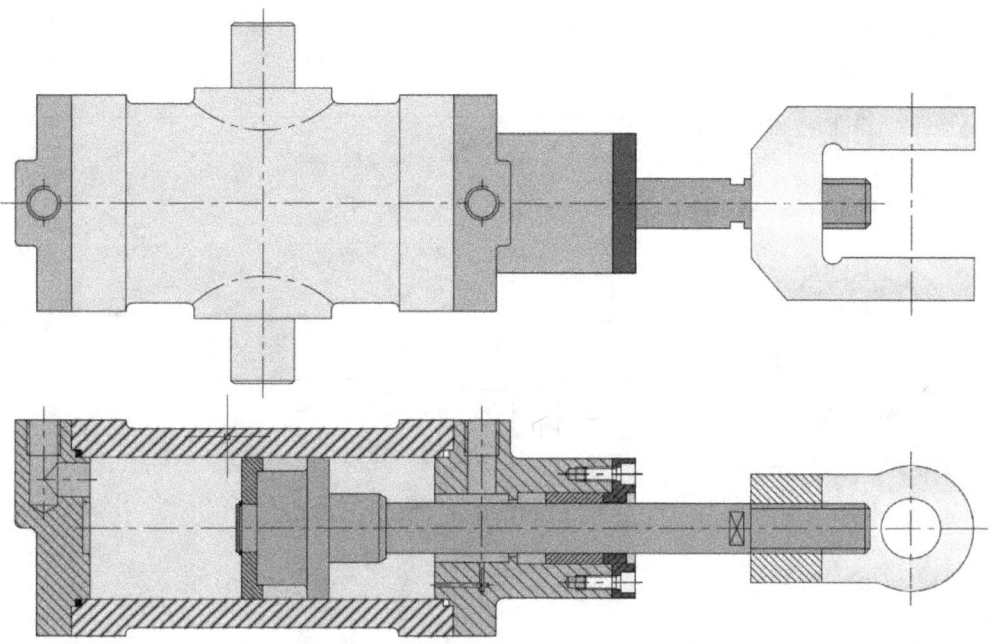

Stripper Cylinder Assembly

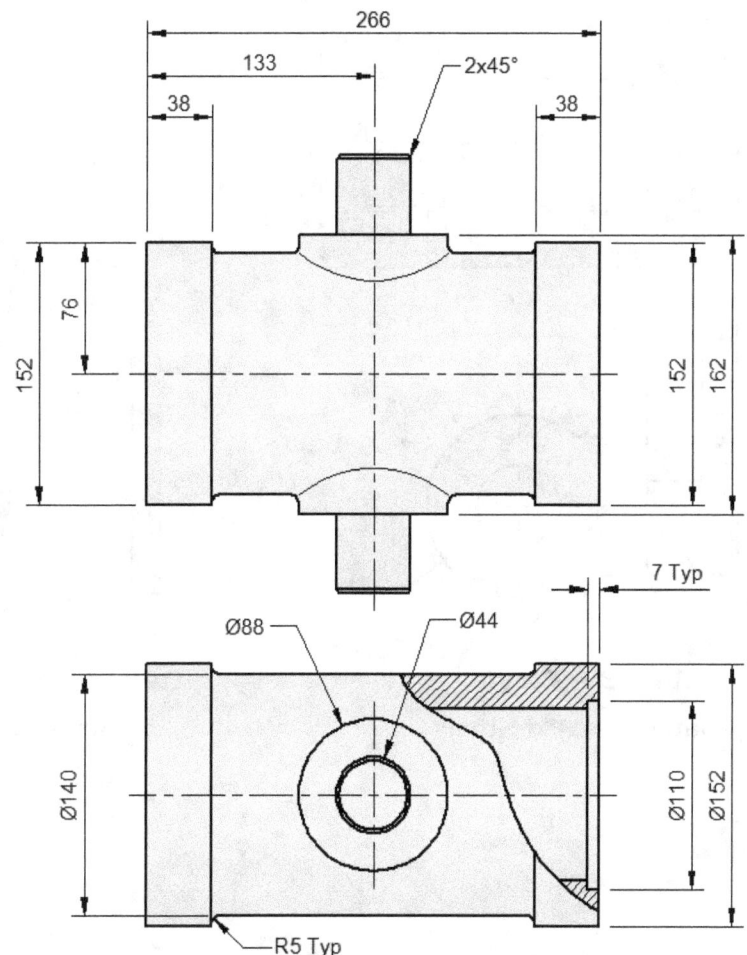

Item 1 - Stripper Cylinder Body
Material – Cast Iron

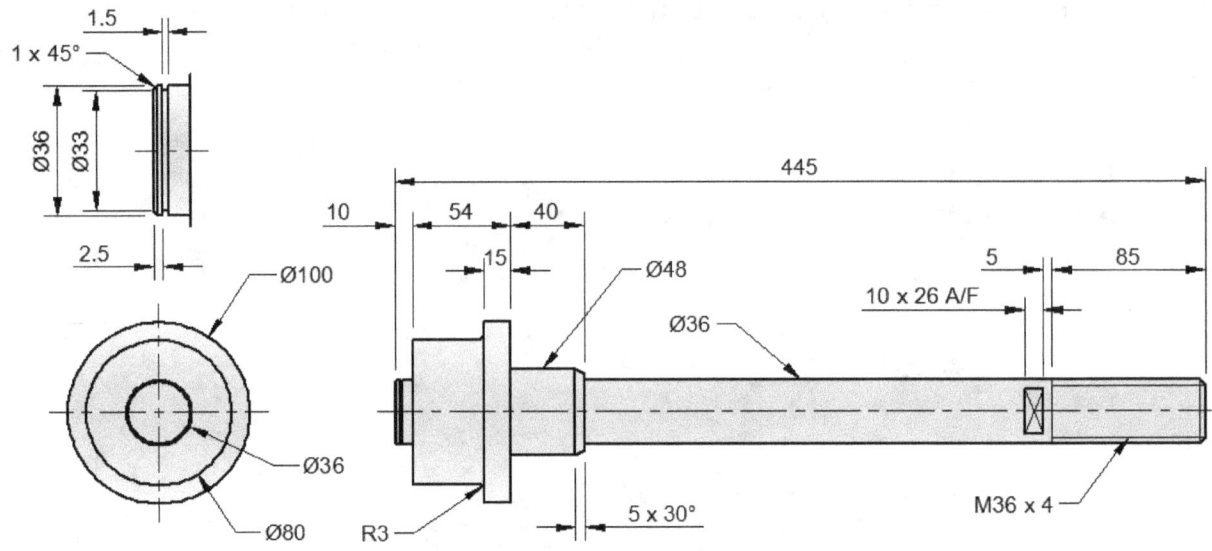

Item 2 – Piston Rod

Material – Mild Steel

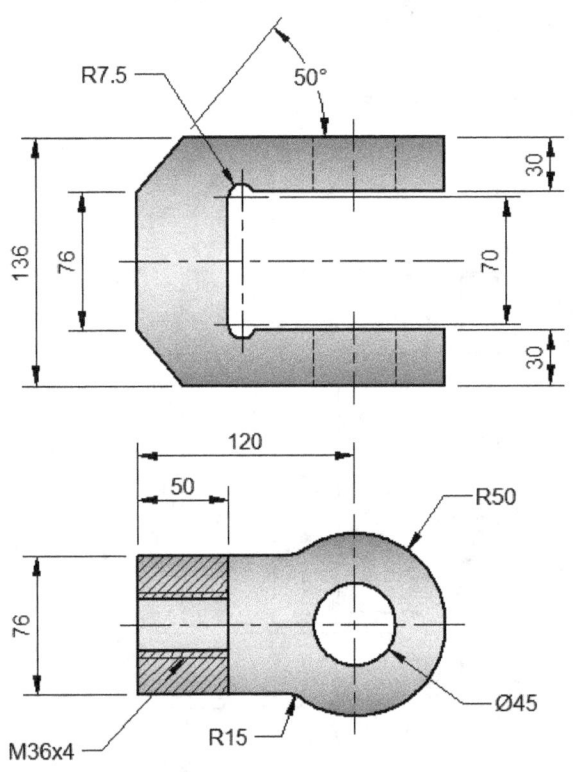

Item 3 – Clevis

Material – Mild Steel

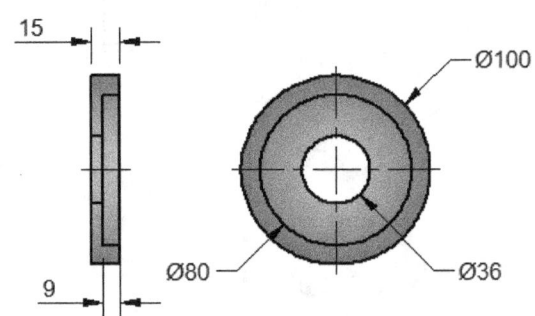

Item 7 – Seal Retaining Cap

Material – Mild Steel

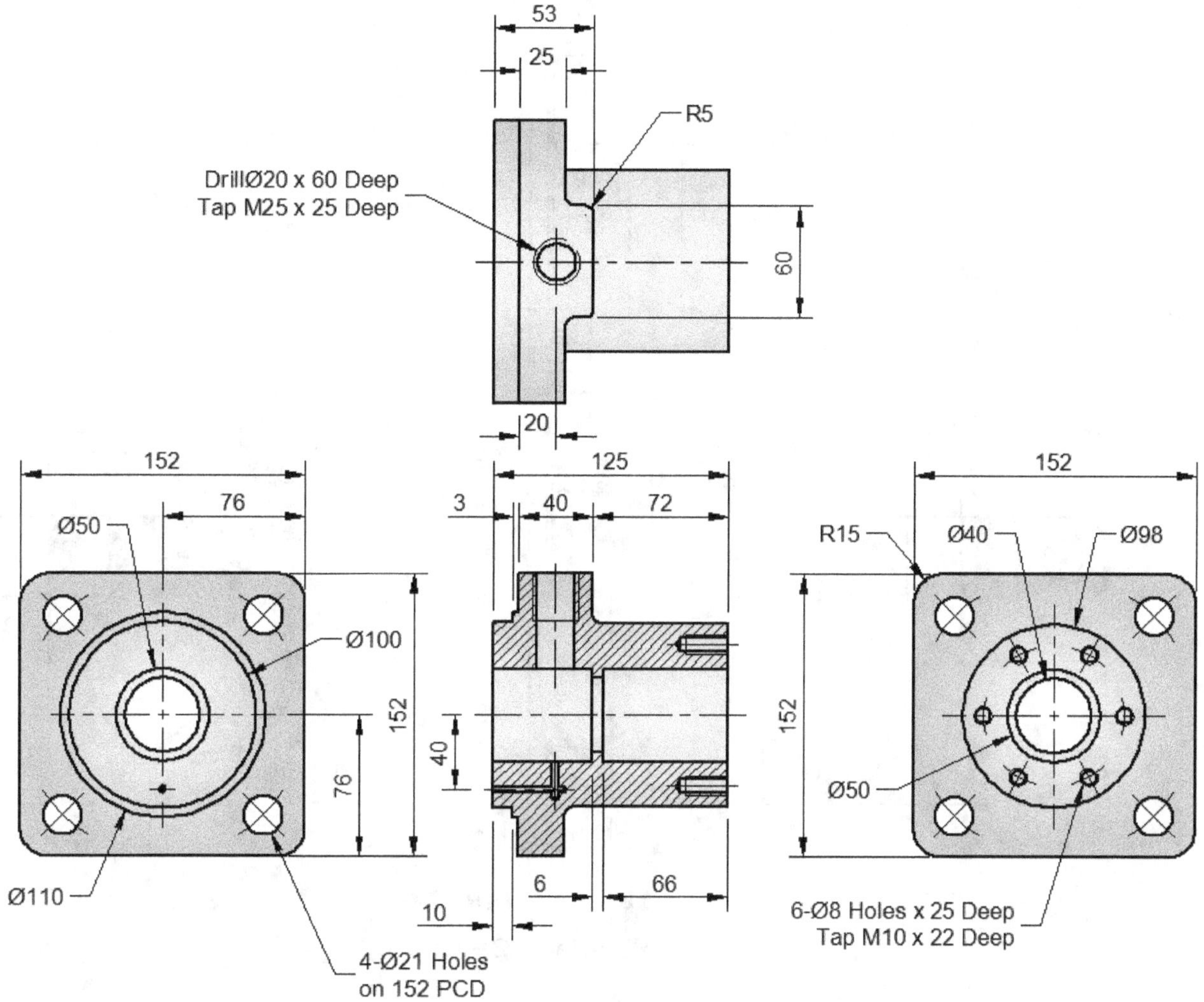

53

25

R5

DrillØ20 x 60 Deep
Tap M25 x 25 Deep

60

20

152

76

Ø50

Ø100

152

76

Ø110

10

4-Ø21 Holes
on 152 PCD

125

3   40   72

40

6   66

152

R15   Ø40   Ø98

Ø50

6-Ø8 Holes x 25 Deep
Tap M10 x 22 Deep

Item 4 – Cylinder Front End
Material – Cast Iron

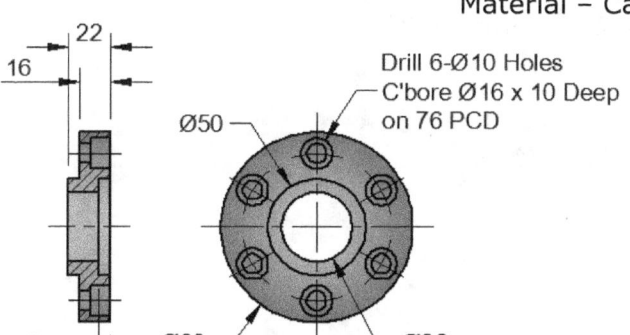

22

16

Ø50

6

Ø98   Ø36

Drill 6-Ø10 Holes
C'bore Ø16 x 10 Deep
on 76 PCD

Item 6 – Seal Cap

Material – Mild Steel

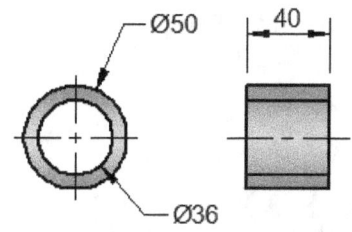

Ø50

40

Ø36

Item 8 – Bush

Material - Bronze

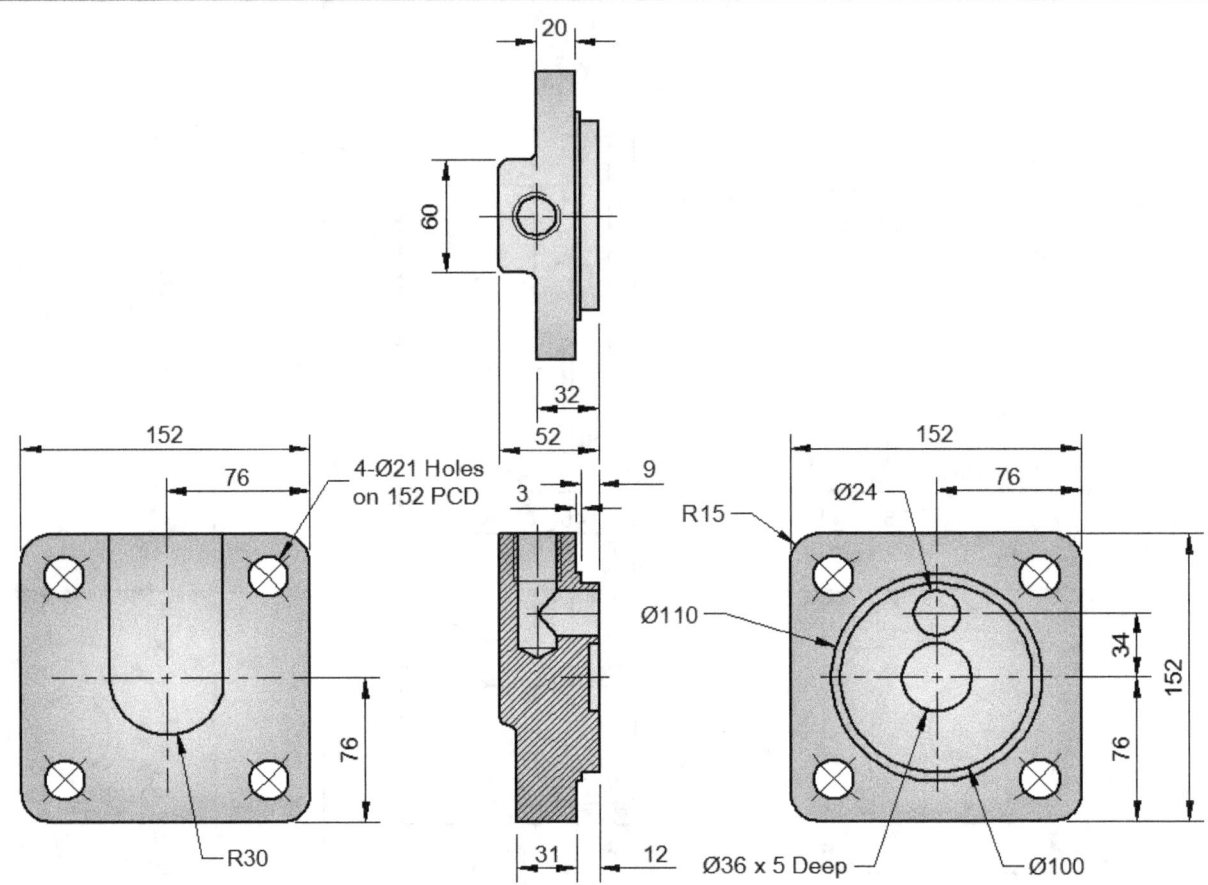

Item 5 – End Cap

Material – Cast Iron

Additional Parts

Socket Head Cap Screws – M10 x 1 x 25 - Commercial

Machine Screws – M20 x 2 x 40 - Commercial

Ø36 External Circlip - Commercial

V Packing - Felt

Skill Practice Exercise MEM09002-SP-0601
Referring to Assembly and Detail drawings above answer the following questions:

1. Which part and item number does the Clevis screw into?

   _____

2. What is the inside diameter of the Stripper Cylinder Body?

   _____

3. Calculate the maximum distance the shaft can move.

   _____

4. How many Machine Screws – M20 x 2? Are required?

   _____

5. Determine the Outside and inside diameter of the V Packing.

   _____

6. Calculate the minimum length of the Stripper Cylinder Assembly.

   _____

7. Name the 2-parts the Bush fits over.

   _____

8. What part does the circlip fir over?

   _____

9. Determine the overall length of the Clevis.

   _____

10. Calculate the distance the Cylinder Front End fits inside the Stripper Cylinder Body.

   _____

Name: _____

Skill Practice Exercise MEM09002-SP-0602

Referring to Assembly and Detail drawings above complete the Material List.

Item Number	Description	Material	Quantity
1			
2			
3			
4			
5			
6			
7			
8			
9			
10			
11			
12			

Name: _____

Topic 7 – Abbreviations, Symbols & Notes:

Required Skills:
- Reading, interpreting information on the drawing.
- Interpret standard abbreviations used in engineering drawings.
- Decipher standard symbols used in engineering drawings.
- Explain standard notes used in engineering drawings.

Required Knowledge:
- Application of AS1100.
- Understanding of the instructions contained in the drawing.
- Symbols used in the drawing.

Abbreviations & Acronyms:

An abbreviation is any shortened form of a word or phrase and an acronym is a form of an abbreviation; in fact, there are three forms of abbreviation, acronyms, initialism, and truncations.

Abbreviations are used on drawings to save time and space. The common abbreviations such as PCD, THD, MAX, MIN, ID & OD are universally understood. Uncommon abbreviations such as PH BRZ (Phosphor Bronze) should be used cautiously because of the possibility of misinterpretation. Abbreviations should conform to the Australian Standards where possible; non-standard abbreviations should be listed on the drawing. List of abbreviations is given in Table 1.

Abbreviations should only be used when their meanings are unquestionably clear to the intended reader; if there is any doubt of confusion, the entire word should be spelt out in full. The same abbreviation should be used for all tenses, the possessive case, participle endings, the singular or plural, and noun and modifying forms.

Acronym:

An acronym is a word formed from the initial parts of a name and can consist of letters or syllables. For example, "Australian and New Zealand Army Corps" is commonly known as ANZAC. We are more familiar with sonar than we are with sound navigation and ranging. SONAR is the more recognizable name used in lieu of "Sound Navigation and Ranging" or SCUBA which stands for "Self-Contained Underwater Breathing Apparatus".

Initialism:

Initialism, or initials, is formed by combining the first letters in a name or expression and each letter is pronounced separately. For example, the "Australian Broadcasting Corporation" is known as the ABC. RFS is the initialism for the Rural Fire Service while DVD is Digital Versatile Disc and ATM, Automated (or Automatic) Teller Machine.

Truncation:

Truncation is the act or process of truncating shortening a word by removing part of it. It can also mean the state of having been truncated. Truncation can involve the removal of the beginning of something, the end of it, the top of it, or another part of it.

In this form of abbreviation, a word is shortened to its first syllable or few letters, for example TUES. is Tuesday and INFO is information.

Jargon:

Jargon is a technical or occupational term developed to help specialists in a specific industry or business communicate quickly and simply with one another. Unfortunately, jargon tends to escape the confines of the narrow fields to which they apply and into our everyday language.

Some examples used in various industries are:

Boatbuilding	Dead-rise	Looking at the hull in cross section, the angle the bottom rises from a horizontal.
	gravo	A piece of timber or ply used to thicken a weak area on the hull or cabin.
	thwart	A seat spanning across a boat, not fore & aft.
Construction	foundation	The earth directly beneath the footing and distributes loads to the strata.
	lath	A metal wire on the frame of a building that serves as a base for laying down stucco or plaster.
	stud	A vertical wall member used to attach other structures, such as walls.
Electrical	capping	A slender plastic or metal channel often used to home cables when fixed to a wall before plastering. A capping accommodates several cables that follow the same route to minimise the use of fittings.
	Daisy chaining	The act of plugging several extension leads into each other which can cause electrical fires.
	neutral	Along with the earth and live comes the neutral wire, a neutral wire completes the entire circuit and directs power back to the station.
Fitting & Machining	apron	The portion of a lathe carriage that contains the clutches, gears, and levers for moving the carriage and protects the mechanism.
	Bastard File	A medium grade or pitch of file for general purposes, especially suitable for mild steel.
	Tool holding	The structure or apparatus designed to hold cutting bit or tools. On a mill, this is often the element that spins, whereas on a lathe, this is the element that is secured to the table.
Mechanical Engineering	allowance	is a planned deviation between an exact dimension and a nominal or theoretical dimension, or between an intermediate-stage dimension and an intended final dimension.
	enthalpy	A measure of the total energy of a thermodynamic system.
	moment	The tendency of a force to cause a rotation about a point or axis which in turn produces bending stresses.
Welding	back Fire	The momentary burning back of a flame into the tip, followed by a snap or pop, then immediate reappearance or burning out of the flame.

| bird poop | Cold, supremely ugly, ropey blobs of metal made by a novice failing a weld. |
| coalescence | The uniting or fusing of metals upon heating. |

Symbols:

Symbols provide a "common language" for drafters all over the world and are intended to communicate design intentions in a clear manner; however, symbols are meaningful only if they are drawn according to relevant standards or conventions. Industry standards have been developed to provide the graphical symbols for the various disciplines. This section describes and illustrates common mechanical, architectural, piping, and electrical symbols.

Unless specified otherwise, the size of dimensioning symbols is consistent with text height. In the majority of cases, all symbols are proportional to the text height.

A list of symbols used in the drafting, welding, structural, piping, electrical and electronic disciplines are shown in Tables 2 to 8.

Table 1 – Safety Symbols:

	Watch you step		Evacuation Assembly Area Sign
	Emergency Exit		Emergency First Aide
	Biohazard		Wear Face Mask
	All Pedestrians		No Smoking
	Danger Construction Site		Fire Extinguisher
	Fire Emergency Telephone		Eye Protection
	Hearing Protection	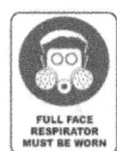	Full Face Respirator
	Foot Protection		Hand Protection
	Asbestos		Do Not Use Ladder
	Hard Hat Area		Safety Harness

Table 2 – Abbreviations:

Fastenings:

A/F	Across Flats	HEX SOC HD	Hexagonal Socket Head
A/P	Across Points	MUSH HD	Mushroom Head
CBORE	Counterbore	RD HD	Round Head
CH HD	Cheese Head	RSD CSK HD	Raised Countersunk Head
CSK	Countersunk	SCR	Screw
CSK HD	Countersunk Head	SOC	Socket
FILL HD	Fillister Head	SQ HD	Square Head
HEX HD	Hexagonal Head		

Materials:

AC	Asbestos Cement	HTS	High Tensile Steel
AL	Aluminium	MI	Malleable Iron
AL AL	Aluminium Alloy	MS	Mild Steel
BRS	Brass	PH BRZ	Phosphorus Bronze
BRZ	Bronze	PVA	Polyvinylacetate
CAD or CD PL	Cadmium Plate	PTFE	Polytetrafluroethylene
CI	Cast Iron	PVC	Polyvinylchloride
CS	Cast Steel	RC	Reinforced Concrete
CRS	Corrosion Resistant Steel	SS	Stainless Steel
GALV	Galvanise	STL	Steel
GALV MS	Galvanised Mild Steel	SPR STL	Spring Steel

Steel Sections:

L	Rolled Steel Angle	PL	Plate
⊏	Rolled Steel Channel	RHS	Rectangular Hollow Section
CHS	Circular Hollow Section	SHS	Square Hollow Section
CRS	Cold Rolled Section	UA	Unequal Angle
EA	Equal Angle	UB	Universal Beam
HRS	Hot Rolled Section	UC	Universal Column

Technical:

ABBR	Abbreviation	DIAG	Diagonal
ABS	Absolute	DIAG	Diagram
ACCEL	Acceleration	DIA	Diameter
AO	Access Opening	ID	Internal Diameter
AP	Access Panel	OD	Outer Diameter
ACC	Accumulator	DP	Diametral Pitch
AR	Acid Resistant	DIM	Dimension
AW	Acid Waste	DC	Direct Current
ACST	Acoustic	DIST	Distance
APC	Acoustic Plaster Ceiling	DWG or DRG	Drawing
ADD	Addendum	EA	Each
AGGR	Aggregate	EQUIV	Equivalent
AIR COND	Air Condition	EST	Estimate
AC	Alternating Current	EXST	Existing
AMDT	Amendment	EXT	External
ANL	Annealed	FIG	Figure
APPROX	Approximate	FLG	Flange

Topic 7 - Abbreviations, Symbols & Notes

ARR or ARRGT	Arrangement	FWD	Forward
AS	Australian Standard	FP	Freezing Point
ASSY	Assembly	FREQ	Frequency
ASSD	Assumed Datum	GA	General Arrangement
AHD	Australian Height Datum	GR	Grade
AUTO	Automatic	HD	Heavy Duty
AUX	Auxiliary	HT or HGT	Height
AV or AVG	Average	HEX	Hexagon
BRG	Bearing	HP	High Pressure
BM	Benchmark	HORIZ	Horizon
M	Bending Moment	HYD	Hydraulic
BLK	Block	INSUL	Insulation or Insulate
BM	Bench Mark	INT	Internal
BRKT	Bracket	ID	Internal Diameter
HB	Brinell Hardness number	SI	International System of Units
BLDG	Building	IP	Intersection Point
CALC	Calculated	LAT HT	Latent Heat
CAP	Capacity	LMC	Least Material Condition
CH	Case Harden	LG	Length
CL	Centreline	LEV	Level
CG	Centre of Gravity	LIQ	Liquid
C/C	Centre-to-Centre	LL	Live Load
CHAM	Chamfer	LONG	Longitudinal
CIRC	Circle	LP	Low Pressure
COL	Column	LUB	Lubricate
COMP	Compression	MACH or M/C	Machine
CONC	Concentric	MK	Mark
CONST	Constant	MATL	Material
CST	Crest	MAX	Maximum
CYL	Cylinder	MMC	Maximum Material Condition
DET	Detail	MECH	Mechanical
MP	Melting Point	SHT	Sheet
MIN	Minimum	SPEC	Specification
MISC	Miscellaneous	SPHER	Spherical
MOD	Modification	SPT	Spigot
E	Modulus of Elasticity	SF	Spot Face
Z	Modulus of Section	SQ	Square
I	Moment of Inertia	STD	Standard
MTG	Mounting	STA	Station
NEG	Negative	STR	Straight
NOM	Nominal	SFL	Structural Floor Level
NTS	Not To Scale	SW	Switch
NO.	Number	SWBD	Switchboard
OCT	Octagon	SYM	Symmetry
OPP	Opposite	TP	Tangent Point
OD	Outside Diameter	TEMP	Temperature
OA	Overall	TBM	Temporary Bench Mark
PH	Phase	TS	Tensile Strength
PCD	Pitch Circle Diameter	THK	Thick
PNEU	Pneumatic	THD	Thread
POS	Position	TOL	Tolerance
POS	Positive	T&G	Tongue & Groove
PRESS	Pressure	XFMR	Transformer
PA	Pressure Angle	TRANSV	Transverse

Topic 7 - Abbreviations, Symbols & Notes

QTY	Quantity	TP	True Position
R or RAD	Radius	TYP	Typical
RECT	Rectangular	ULT	Ultimate
RL	Reduced Level	UTS	Ultimate Tensile Strength
REF	Reference	UCUT	Undercut
RM	Reference Mark	U.N.O.	Unless Noted Otherwise
RFS	Regardless of Feature Size	UBP	Universal Bearing Pile
REINF	Reinforcement	VAC	Vacuum
REQD	Required	VERT	Vertical
REV	Revision	VH	Vickers Hardness
RH	Right Hand	VOL	Volume
RHA	Rockwell Hardness A	WG	Water Gauge
RHB	Rockwell Hardness B	WL	Waterline, Water Level
RHC	Rockwell Hardness C	WB	Weatherboard
R_a	Roughness Value	W	Wide
RD	Round	WL	Wind Load
SCHED	Schedule	W/O	Without
SECT	Section	YP	Yield Point

Table 3 - Common Engineering Drawing Symbols:

Datum Identifier	▲	
Diameter	Ø	Ø18 Ø9
Equal	=	
Feature Identification	▭	60 15 15 30 15
Projection – First Angle	◁⊕	1 2 3 A B
Projection – Third Angle	⊕◁	
Radius	R	R20 R10
Slope	◁	1:4
Square	□	□8
Taper	▷	1:2.5
Counterbore	⊔	
Countersink	⊔	
Depth	⋁	

Table 4 - Welding Symbols:

Basic Gas & Arc Welding

Fillet	
Bead	
General Butt	
Square Butt	
Single Bevel Butt	
Single Vee Butt	
Single 'U' Butt	
Single 'J' Butt	
Plug or Slot	
Stud	
Surfacing	

Resistance Welding

Spot	
Seam	
Mash Seam	
Stitch	
Mash Stitch	
Projection	
Flash Butt	
Resistance Butt	

Resistance Supplementary Symbols

All Round Weld	
Flush Contour	

Gas & Arc Supplementary Symbols

All Round Weld	
Field or On-Site Weld	
Backing Strip or Bar	

Symbolic Representation

Fillet Located on the Arrow Side	
Fillet Located on the Other Side	
Fillet Located on Both Sides	
Vee Butt Located on the Arrow Side	
Vee Butt Located on the Other Side	
Vee Butt Located on the Both Sides	

Contour Symbols

Flush Weld	
Convex Weld	
Concave Weld	

Table 5 – Structural Steel Sections:

Section	Diagram	Abbr.	Example
Universal Beam		UB	610 UB 125 Where: 610 = Depth (Nominal) 125 = mass in kg/m
Universal Column		UC	310 UC 158 Where: 310 = Depth (Nominal) 158 = mass in kg/m
Tapered Flange Beam		TFB	125 TFB Where: 125 = Depth
Parallel Flange Channel		PFC	380 PFC Where: 380 = Depth
Tapered Flange Channel		TFC	125 TFC 13.4 Where: 125 = Depth 13.4 = mass in kg/m
Equal Angle		EA	100 x 100 x 10 EA Where: 100 = Depth 100 = Breadth t = Thickness
Unequal Angle		UA	100 x 75 x 8 UA Where: 100 = Depth 75 = Breadth 8 = Thickness

Table 6 – Structural Steel Profiles:

Profile		Abbr.	Description

Square Hollow Section		SHS	100 x 100 x 6 SHS
			Where:
			   100 = Depth
			   100 = Breadth
			      6 = Thickness

Rectangular Hollow Section		RHS	125 x 75 x 6 RHS
			Where:
			   125 = Depth
			     75 = Breadth
			       6 = Thickness

Circular Hollow Section		CHS	75 OD x 3 CHS
			Where:
			   75 = Outside Diameter
			     t = Thickness

Round Bar & Rod		RD	50 RD Where: 50 = Diameter

Square Bar		SQ	25 SQ
			Where:
			   25 = Depth
			   25 = Breadth

Flat Bar		FL	75 x 10 FL
			Where:
			   75 = Width
			   10 = Thickness

Plate		PL	900 x 10 PL x 2400
			Where:
			     900 = Width
			       10 = Thickness
			   2400 = Length

Table 7 – Pipeline Symbols:

Valves:

Valve – General Symbol		Non-Return Valve	
Globe Valve		Three Way Valve	
Globe Valve with Maximum Flow Adjustment		Angle Valve	
Ball Valve		Reducing Valve	
Butterfly Valve	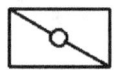	Fire Hydrant	
Strainer	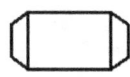	Suction Pipe Strainer	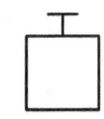
Flow Meter – Recording	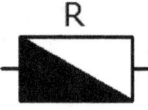	Flow Metre – Non Recording	
Steam Trap			

Operation:

Hand		Diaphragm	
Solenoid		Lock Shield	
Spring		Float	
Counterweight			

Gauges:

Thermostat		Humidistat	
Thermometer Dial		Pressure Gauge	
Sight Glass			

Miscellaneous:

Centrifugal Pump – Solid Casing		Centrifugal Pump – Split Casing	
Injector or Ejector		Coil – Heating or Cooling	
Pipe Crossing		Vertical Pipe	
Vertical Pipe Rising		Vertical Pipe Dropping	
Direction of Flow		Rise in the Direction of the Flow	
Fall in the Direction of the Flow		"T" Piece	
90° Elbow		45° Elbow	
Flanged Connection		Blank Flange	
Union Joint		Anchor	
Hanger		Orifice Plate	
Open Vent		Anti-Convection Loop	
Pipe Guide		Air Cock	
Automatic Air Cock		Air Vessel	
Tundish		Expansion Bellows	
Expansion Bend		Expansion Bend – Lyre Bend	
Strainer – "Y" Type		Strainer – Line Type	
Flexible Pipe (Hose)			

Table 8 - Mechanical Symbols:

Indication of Surface Texture:
Surface Texture is indicated on a drawing as shown in the following image:

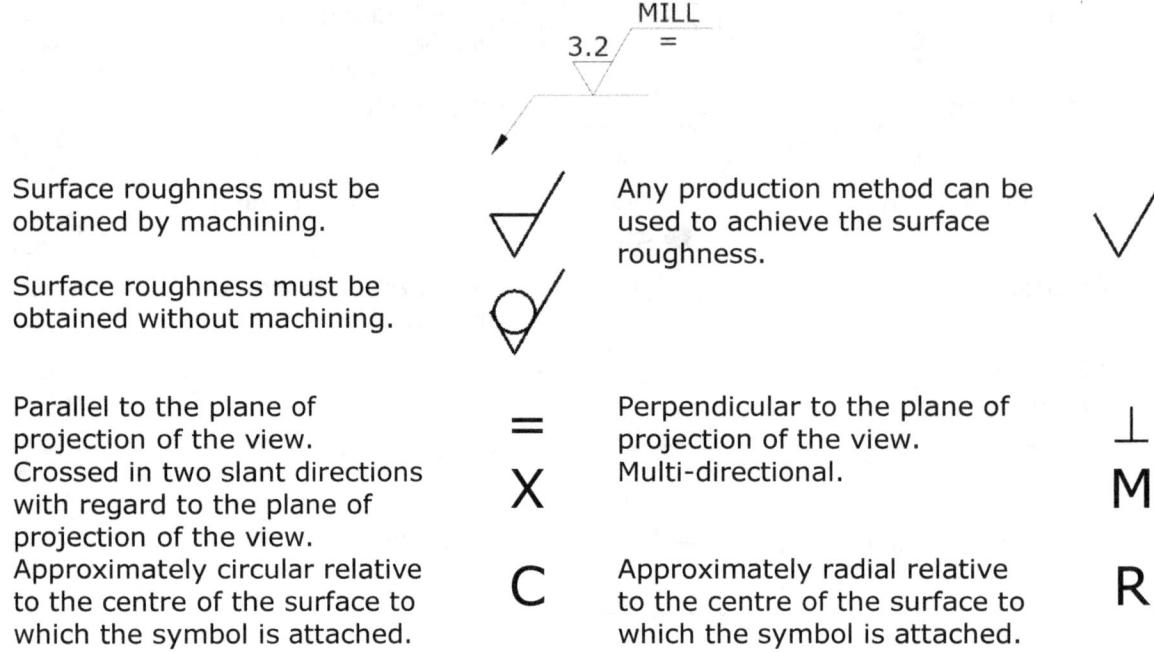

Surface roughness must be obtained by machining. Surface roughness must be obtained without machining.		Any production method can be used to achieve the surface roughness.	
Parallel to the plane of projection of the view. Crossed in two slant directions with regard to the plane of projection of the view.	= X	Perpendicular to the plane of projection of the view. Multi-directional.	⊥ M
Approximately circular relative to the centre of the surface to which the symbol is attached.	C	Approximately radial relative to the centre of the surface to which the symbol is attached.	R

Geometric Tolerance:
Geometric Tolerance is applied to the geometry or shape of the component/s through a special symbol shown in the following image: Typical examples are:
- 2 holes must be concentric to each other.
- A vertical surface must be perpendicular to a datum surface.
- The selected surface must be flat.

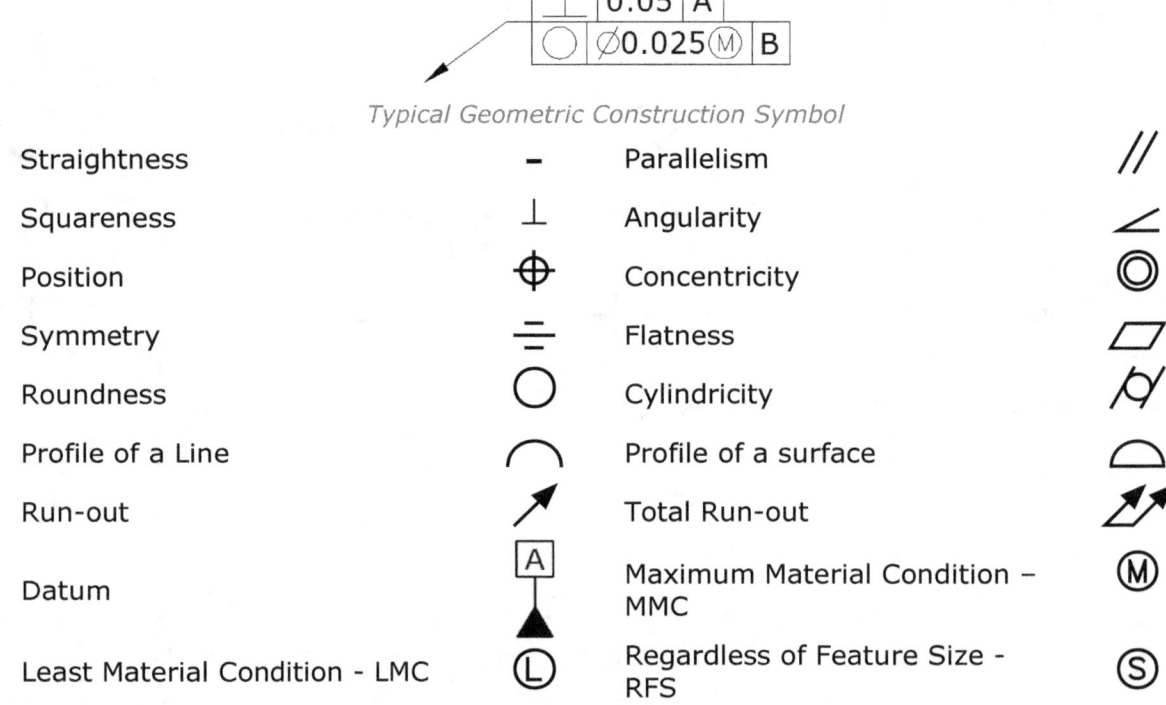

Typical Geometric Construction Symbol

Straightness	–	Parallelism	//
Squareness	⊥	Angularity	∠
Position	⊕	Concentricity	◎
Symmetry	ⲛ	Flatness	▱
Roundness	O	Cylindricity	⌭
Profile of a Line	⌒	Profile of a surface	⌓
Run-out	↗	Total Run-out	↗↗
Datum	[A]▲	Maximum Material Condition – MMC	Ⓜ
Least Material Condition - LMC	Ⓛ	Regardless of Feature Size - RFS	Ⓢ

Table 9 – Electrical Symbols:

Luminaires & Domestic Appliances:

Luminaire – General Symbol		Luminaire fixed to wall	
Luminaire with number and power of lamps in the group	5 x 40 W	Luminaire with built-in switch	
Emergency Lighting Luminaire		Signal Lamp	
Warning alarm, Panic light		Spotlight	
Floodlight		Lamp with reflector	
Fluorescent – Single tube		Fluorescent – Double tubes	
Fluorescent – Triple tubes		Fluorescent - Alternate Symbol	2 x 40 W
Discharge Lamp		Auxiliary apparatus for discharging lamp.	
Electrical Appliance – Basic symbol		Electric Range	R
Garbage Disposal Unit	GD	Exhaust Fan	EF
Air Conditioner	AC	Electric Heater	H
Fan Heater	FH	Electric Heater – Alternate symbol	

Distribution Boards:

Main Switchboard	MSB	Meter Board	MB
Distribution Board	DSB	Automatic Telephone Exchange	PABX
Fire Indicator Board	FIB		

Switches & Buttons:

One Way Switches – Single, Two & Three Poles		Single Pole Pull Switch	
Light Dimmer Switch with variable control		Multi-Position Switch for different degrees of lighting	
Two-Way Switch		Intermediate Switch	
Time Switch		Push Button	
Luminous Push Button		Manually Operated Fire Alarm	

Restricted Access Push Button		Remote Controlled Equipment	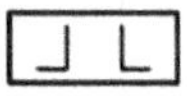

Socket Outlets:

General Symbol		Multiple Socket Outlet – 'n' for plugs	
Switched Socket Outlet		Socket Outlet with Protective Earth Contact	
Single Phase Socket Switched and Earthed		Socket Outlet with Protective Interlocking Switch	
Multi			
-Phase Socket Outlet			

Miscellaneous:

Point of Attachment	POA	Earth	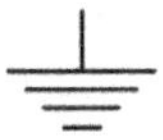
Battery		Lightning Arrestor	

Telecommunications, Radio & Television Apparatus:

General Symbol - Telecommunications	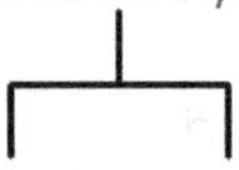	Television	TV
Radio	R	Sound	S
Aerial – Antenna		Loudspeaker	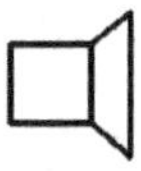
Radio Receiving Set	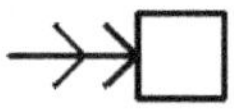	Amplifying Equipment	
Microphone	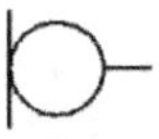	Telephone Outlet – Wall Mounted	
Telephone Installed on Wall		Telephone Outlet - Floor	
Telephone Installed on Floor		Intercom	H
Through Switchboard	S	Direct Line	D
Distribution Point			

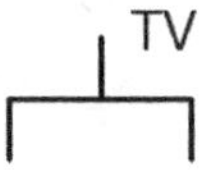

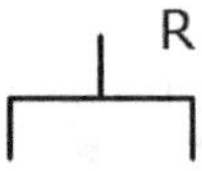

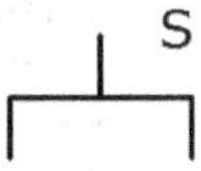

Miscellaneous Apparatus and Appliances:

Thermal Fire Alarm Detector Head	◑	Motor	Ⓖ
Generator	Ⓜ	Ceiling Fan	⊘
Rectifying Unit DC Power Supply	⊳	Electric Bell	⛑
Electric Buzzer	⊽	Siren	⇧
Horn	⊏▭⟍	Clock	⊕

Cable Codes:

Electric Power	E	Telephony	F
Data Circuit	T	Video Circuit	V
Audio Circuit	S	Lighting	L
Street Lighting	SL		

Table 10 – Electronic Symbols:

Indicating Instruments:

Ammeter	(A)	Voltmeter	(V)
Frequency Meter	(Hz)		

Contacts for Switches & Relays:

Make Contact	─o o─	Break Contact	─o o─

Switchgear:

Circuit Breaker		Make Contactor	
Break Contactor		Contactor with Coil Type Blow-Out Device	

Coils for Telephone Type Relays:

General Symbol – Relay Coil		Relay Coil with 1300 ohm Winding	1300

Contact Units for Telephone:

Make Contact Unit		Break Contact Unit	
Changeover Contact Unit (break before make)		Changeover Contact Unit (make before break)	

Diode Devices:

General Symbol – Preferred		General Symbol – Alternate	
Tunnel Diode		Thyristor	
Reverse Blocking Triode Thyristor – 'n' gate, Anode controlled		Reverse Blocking Triode Thyristor – 'p' gate, Cathode controlled	
pnp Transistor (also pnip transistor if omission of the intrinsic region will not result in ambiguity)		npn Transistor with collector connected to envelope	
Unijunction Transistor with 'p' type base			

Earth and Frame Connections:

General Symbol – Earth or Ground		Protective Earth	
Noiseless or Clean Earth Connection		Earth Connection	

Topic 7 - Abbreviations, Symbols & Notes

Miscellaneous:

Direct Current or Steady Current – Preferred		Direct Current or Steady Current – Alternative	
Alternating Current		Conductor or Group of Condensors	
Positive Polarity		Negative Polarity	
Flexible Conductor		Unconnected Cable or Conductor	
Unconnected Cable or Conductor Especially Insulated		Jumper	
Two Conductors	or	Three Conductors	or
'n' Conductors	n	Envelope (Tank)	
Boundary Line		Permanent Magnet	
Fault		Indicator	
Hot Cathode – Preferred		Hot Cathode – Alternate	
Photoelectric Cathode		Anode (Plate) or Collector	
Brush on Slip-Ring		Brush on Communicator	

Skill Practice Exercises:

Skill Practice Exercise MEM09002-SP-0701

Complete the following by adding the abbreviation or the expanded form of the abbreviation.

	Modulus of Elasticity	FILL HD	
	Reference	SQ	
CRS			Specification
FREQ			Rolled Steel Angle
	High Tensile Steel	CHS	
PCD			Socket
	Galvanised Mild Steel	HYD	
BM			Hexagonal Head
	Bronze	SPR STL	
UCUT			Required
DIA			Equivalent
	Miscellaneous	CBORE	
	Addendum	AL AL	
AP			Modulus of Inertia
CL			Across Flats

Name: _____

Skill Practice Exercise MEM09002-SP-0702

Complete the following by adding the name or the symbol of the feature.

Ø			Surface must be obtained without machining
⋈		⌣⌣⌣	
	Hanger	⏚	
	Feature Identification	⊕▷	
∧		⚡	
	Main Switchboard		Loudspeaker
	All Round Weld	⟋	
⊣×			Circuit Breaker
	Float Operated	⊠	
▽			Slope
▢▢	Datum	⊻	
◯ 5 x 40 W			Spot Weld
	Push Button		Counterbore
⊢──┤		⊕ ◁	
▷		⏾	
	Spotlight		Fillet Weld
Ⓥ		⚑	
	Tunnel Diode		First Angle Projection
	Permanent Magnet	▢	

Name: _____

Practice Competency Test

Referring to drawing MKD-DDg-013 Sheets 1 to 4 answer the following questions.

Sheet 1:

1. How many items are required in manufacturing the assembly?

   _____

2. What projection has been produced to produce the drawing?

   _____

3. What was modified on the drawing to cause a new revision?

   _____

4. Give the full name of the material used to manufacture the Jamb Pad?

   _____

5. To what standard was the drawing produced?

   _____

6. What do the 2 short curved lines represent on the Lever in the End View?

   _____

7. If the assembly was to be transported with the Lever in the horizontal position, calculate the total length of the assembly.

   _____

8. What is the total number of components that is required for the assembly?

   _____

9. Which view does not show the width of the assembly?

   _____

10. What is the last step in the assembly before transportation?

   _____

Name: _____

Sheet 2:

11. What is the height of the Ø16 hole above the datum surface?

_____

12. What is the internal radius of the slots?

_____

13. What type of view is shown between the Sectioned Front and Top Views?

_____

14. How many tapped holes are required Base?

_____

15. What is the of line type name for the line connecting the symbols A in the Top View?

_____

16. What is the overall length of the Base?

_____

17. Who drew the original drawing and on what date was it drawn?

_____

Name: _____

Sheet 3:

18. How many times has the drawing been reissued?

_____

19. What does the stroke under the R26 dimension indicate?

_____

20. What do the brackets around the 120 dimension indicate?

_____

21. What is the eccentricity between the R26 outline and Ø16 hole?

_____

22. What change forced a reissue of the drawing?

_____

23. What is the name of the firm who produced the drawing?

_____

24. What is the diameter of the handle of the Lever Arm?

_____

25. What do the 2-broken lines indicate in the Top View?

_____

26. What is the size of the Sheet?

_____

Name: _____

Sheet 4:

27. What orientation is the sheet?

_____

28. What is the drill size for the M12 tapped hole?

_____

29. What does the abbreviation C'BORE mean?

_____

30. What is the thickness of the Splinter Pad?

_____

31. Who checked the drawing?

_____

32. What are the centre-to-centre distances between the 3-10.5 holes?

_____

Name: _____

General Questions:

33. What does P.C.D. mean?

_____

34. Write the abbreviation for assembly?

_____

35. Arrowheads, Dimension Line, Leader Line, Reference Line, Dimension Text Height and Gap are all features of a dimension; name the other line found in a dimension?

_____

36. From the images below, which represents an isometric drawing?

A.   B.   C.   D.

_____

37. What does the abbreviation EQUIV mean?

_____

38. A single thick line around the edge of a sheet is one method of indicating the border; name the other method of border system.

_____

39. What is the purpose of a Revision Block on a drawing?

_____

Name: _____

40. Identify the names of the lettered line styles?

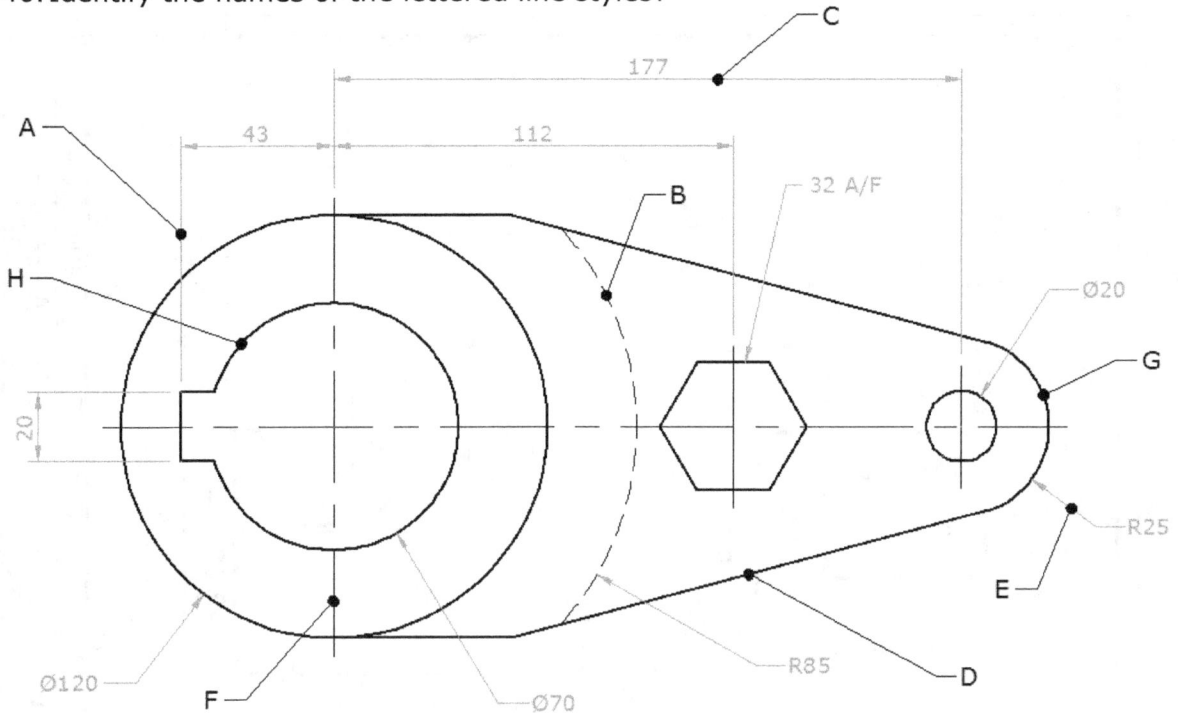

A. _____   B. _____

C. _____   D. _____

E. _____   F. _____

G. _____   H. _____

41. Isometric, Axonometric, Perspective and Oblique are what types of drawings?

_____

Name: _____

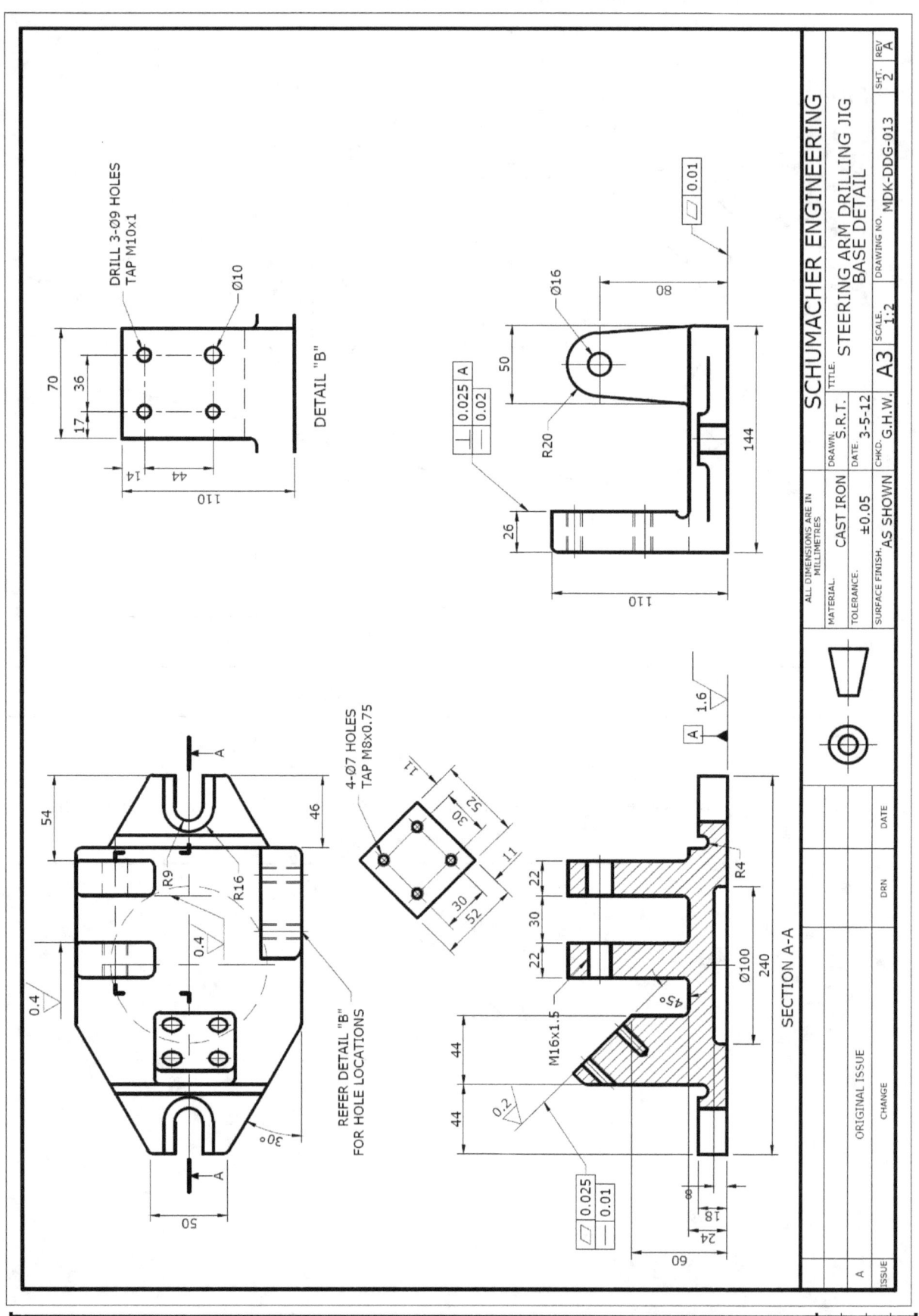

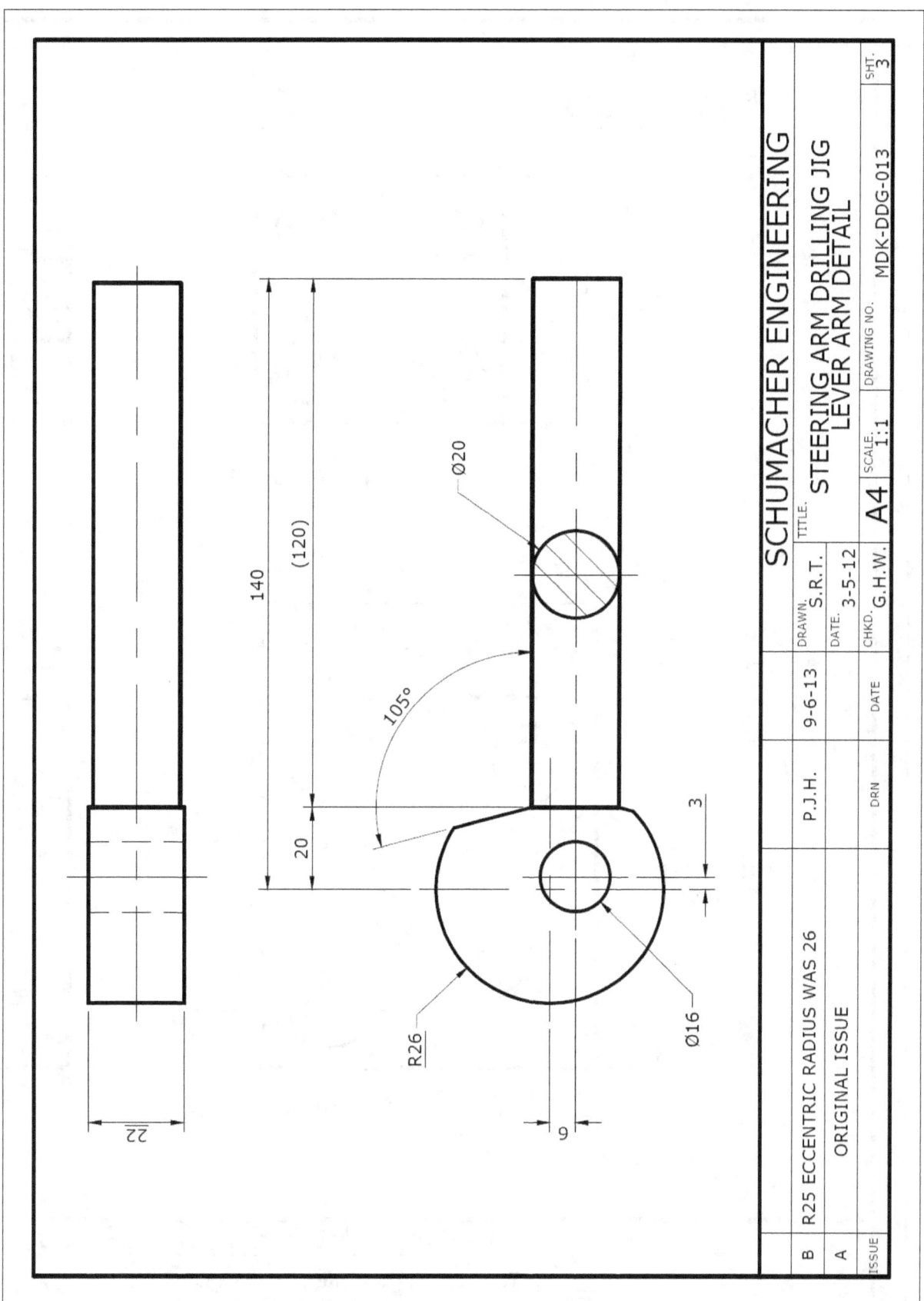

SCHUMACHER ENGINEERING

TITLE. STEERING ARM DRILLING JIG
LEVER ARM DETAIL

DRAWN. S.R.T.

DATE. 3-5-12

CHKD. G.H.W.

A4 | SCALE: 1:1 | DRAWING NO. MDK-DDG-013 | SHT. 3

ISSUE		DRN	DATE
B	R25 ECCENTRIC RADIUS WAS 26	P.J.H.	9-6-13
A	ORIGINAL ISSUE		

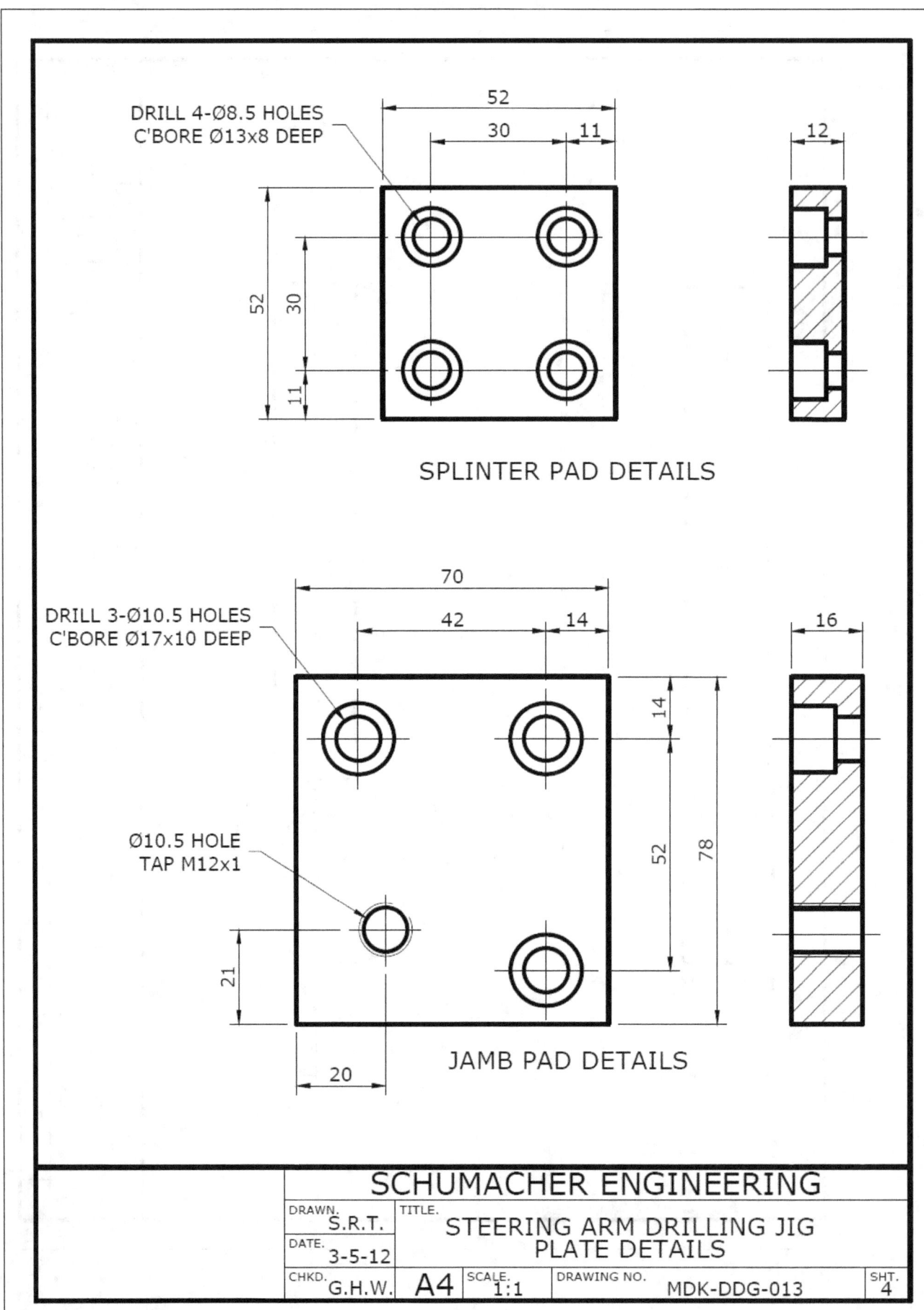

DRILL 4-Ø8.5 HOLES
C'BORE Ø13x8 DEEP

52
30
11
52
30
11
12

SPLINTER PAD DETAILS

DRILL 3-Ø10.5 HOLES
C'BORE Ø17x10 DEEP

70
42
14
14
16

Ø10.5 HOLE
TAP M12x1

52
78
21
20

JAMB PAD DETAILS

	SCHUMACHER ENGINEERING			
DRAWN. S.R.T.	TITLE. STEERING ARM DRILLING JIG PLATE DETAILS			
DATE. 3-5-12				
CHKD. G.H.W.	A4	SCALE. 1:1	DRAWING NO. MDK-DDG-013	SHT. 4

www.ingramcontent.com/pod-product-compliance
Lightning Source LLC
Chambersburg PA
CBHW081811220526
45467CB00006B/2162